Fluid Mechanics Exam File

Stanley Klemetson, Brigham Young University, Editor; Nazeer Ahmed, University of Louisville; L. Carlos Diaz, University of Pittsburgh at Johnstown; John Giger, Rose State College; M. R. Hansen, Oregon Institute of Technology; S. Graham Kelly, The University of Akron; Robert Kern, Milwaukee School of Engineering; Erol Kilik, California State University, Long Beach; Edgar O'Rear, III, The University of Oklahoma; Paul Stymelski, California State Polytechnic University, Pomona; Edward Wetzel, Lafayette College; Timothy Zeigler, Southern Illinois University at Carbondale

ENGINEERING PRESS, INC. SAN JOSE, CALIFORNIA 95103-0001

Donald G. Newnan, Ph.D.
Exam File Series Editor

Printed in the United States of America

Library of Congress Cataloging in Publication Data
Main entry under title:

Fluid mechanics exam file.

1. Fluid mechanics--Problems, exercises, etc.
I. Klemetson, Stanley L.
TA357.F6836 1985 620.1'06'076 84-24693
ISBN 0-910554-48-X

5 4 3 2 1

Engineering Press, Inc. P.O. Box 1 San Jose, Calfornia 95103-0001

Contents

Foreword v

1. FLUID PROPERTIES
   - Mass and Weight 1
   - Compressibility 3
   - Perfect Gases 4
   - Surface Tension 7

2. FLUID STATICS
   - Absolute and Gage Pressure 9
   - Manometers 13
   - Forces on Surfaces 20
   - Buoyancy and Stability 31
   - Acceleration on Fluid Masses 40

3. FLUIDS IN MOTION
   - Velocity Distribution and Rate of Flow 49
   - Continuity Equation 54
   - Bernoulli's Equation 64

4. MOMENTUM PRINCIPLE
   - Forces in Pipes 73
   - Linear and Angular Momentum 83

5. ENERGY PRINCIPLE
   - Applications of Bernoulli's and Energy Equations 91
   - Energy Grade Line 96
   - Pumps and Turbines in the System 97

6. DIMENSIONAL ANALYSIS
   - Buckingham Pi Theorem 107
   - Dimensionless Numbers 109
   - Similitude and Model Analysis 110
   - Turbomachinery's Dimension Analysis 116

7. VISCOSITY
   - Measurement of Viscosity 117
   - Temperature Effects 122
   - Reynold's Number 123

8. SURFACE RESISTANCE AND BOUNDARY LAYER
Laminar and Turbulent Flows 125
Darcy's Equation 130
Moody Diagram and Roughness 132

9. FLOW IN CONDUITS
Friction Factor and Losses in Pipes 137
Minor Losses 146
Pumps and Turbines in Real Systems 153

10. CONDUIT SYSTEMS
Pipes in Series 159
Pipes in Parallel 163
Non-Circular Conduits 171
Equivalent Length 172
Pipe Networks 175

11. DRAG AND LIFT
Vortex from Cylindrical Bodies 179
Friction and Pressure Drag 180
Drag in Three Dimensions 182
Lift 183

12. COMPRESSIBLE FLOW
Mach Number and Critical Conditions 185
Shock Waves and Water Hammer 187

13. FLOW MEASUREMENTS
Velocity and Pressure: Pitot and Current Meters 189
Flow Rate: Venturi and Weirs 195
Measurements of Compressible Flow 200

14. TURBOMACHINERY
Pumps 201
Specific Speed 202
Suction Design 203

15. OPEN CHANNEL FLOW
Energy Relations 205
Channel Design 207
Uniform Flow 210
Hydraulic Jump 217

Foreword

A common request by students in most engineering courses is to have solutions for the problems they are working so that it will be easier to check their approach to the problems. The more exposure that a student has to a variety of problems and problem statements in fluid mechanics the easier it becomes to quickly set up the approach to the problem and find the solution. When it is time to study for an exam, the problems in most textbooks and the student's own lecture notes provide a starting point for the studying process. However, neither of these sources place the student in the test situation with an opportunity to also check his or her own answers. Professors often save their old exams and pick the best of the problems for future exams. In most cases these old exams would provide a method for studying for an exam, but are unavailable to the student.

Professors from many areas of the country have agreed to open their exam files and allow the problems and solutions to be published. These professors teach elementary fluid mechanics and hydraulics from the many books that are published on the subject. The problems in this publication were carefully selected to cover the fundamentals of fluid mechanics found in these textbooks. In general, the problems are just as they appeared in an actual course examination. Similarly, the solutions are those prepared by the professor who wrote the exam problem. Thus this book is a combination of authentic examination problems, together with the professors' own solutions.

Some of the subject areas are approached slightly differently by various professors and textbooks. The differences that you should watch out for are related to the choice of engineering gas constant values, form of Bernoulli's equation, and labeling on the Moody diagram. You should be able to obtain the correct answers using any of the methods currently being used. It is common to have data values in problem statements that are not consistent with the rules for significant numbers. You should assume that any number with less than three significant figures is an exact number for all of the calculations.

I know that you will enjoy using *Fluid Mechanics Exam File* since it is almost like having your own answer book for your textbook. If you find any errors in the problems we would appreciate it if you would write the Series Editor, at the Engineering Press address. I hope that you find this material helps you in gaining a better understanding of fluid mechanics and in doing better on your exams.

Stanley L. Klemetson
Editor

1 FLUID PROPERTIES

MASS AND WEIGHT

1-1

While comparing masses with the help of a mechanical balance, it was found out that mercury occupies a volume of 3 ft^3 with a given mass of 78.689 slug at 68^0F. Determine the density and weight per unit volume of mercury at the given temperature.

****

$$\text{Density} = \rho = \frac{\text{mass}}{\text{volume}} = \frac{M}{V} = \frac{78.689}{3}$$

or, $\rho = 26.230$ slug/ft^3

weight per unit volume $= \gamma = \rho g$, where g is gravitational acceleration with a value of 32.174 ft/sec^2.

or, $\gamma = 26.230 \times 32.174 = 843.92$ lbf/ft^3

1-2

A 8 cm diameter by 2.5 cm thick hockey puck weighs 3N. Determine its density (kg/m^3) and specific gravity.

****

$$\forall = \frac{\pi(.08)^2}{4}(.025) = 1.26 * 10^{-4}\ m^3$$

$$\rho = \frac{\gamma}{g} = \frac{Wt.}{g\forall} = \frac{3\ N}{9.8\ \frac{m}{s^2} * 1.26 * 10^{-4}\ m^3}$$

$$\rho = 2430\ kg/m^3$$

$$S = \frac{2430\ kg/m^3}{1000\ kg/m^3} = 2.43$$

1-3

A fluid has a specific gravity of 1.26. Compute the weight in kN of a 5.0 cubic meter volume of the fluid.

****

$$sg_{fluid} = \frac{\gamma_{fluid}}{\gamma_{water}} \qquad \gamma_{fluid} = (\gamma_{water})(sg_{fluid})$$

$$\gamma_{fluid} = \left(9.81\ \frac{kN}{m^3}\right)(1.26) = 12.36\ \frac{kN}{m^3}$$

$$Weight_{fluid} = (\gamma_{fluid})(Volume_{fluid})$$

$$= \left(12.36\ \frac{kN}{m^3}\right)(5.0\ m^3)$$

$$Weight_{fluid} = 61.8\ kN$$

1-4

The specific volume of a liquid is 0.0011 m^3/kg. Determine the specific gravity of the liquid.

****

$$\rho = \frac{1}{v} = \frac{1}{.0011\ m^3/Kg} = 909.1\ Kg/m^3$$

$$s.g. = \frac{\rho}{\rho_{water}} = \frac{909.1\ Kg/m^3}{1000\ Kg/m^3} \Rightarrow \underline{\underline{s.g. = 0.91}}$$

COMPRESSIBILITY

1-5

Show that for a perfect gas, the bulk modulus of elasticity is $E_v = k \cdot p$ in an isentropic process and $E_v = p$ in an isothermal process. Note: k is the ratio of specific heats.

****

Bulk modulus of elasticity is given by $E_v = \frac{dp}{(d\rho/\rho)}$ (*)

For an isentropic process $p = C.\rho^k$ where C is a constant.

Differentiating this relation and substitution into (*) eqn. give

$$E_v = \frac{C.k.\rho^{k-1}.d\rho}{(d\rho/\rho)} = C.k.\rho^k = k.p$$

For a perfect gas $p = \rho RT$

Differentiating gives : $dp = R(d\rho.T + \rho.dT)$

For an isothermal process : $dp = R.T.d\rho$

Therefore : $E_v = \frac{R.T.d\rho}{d\rho/\rho} = R.T.\rho = p$

PERFECT GASES

1-6

Prove that for a perfect gas, enthalpy is a function of temperature only. Also, show that $C_p - C_v = R$ where R is the gas constant C_p and C_v are specific heats at constant pressure and temperature, respectively.

****

Enthalpy is defined as $h = u + p\forall$

For an ideal gas $du = C_v . dT$

$p\forall = RT$ can be written.

Therefore, $h = u(T) + RT = h(T)$: function of temp. only.

Now, let's differentiate this expression:

$$dh = du + R . dT$$

For a perfect gas, $dh = C_p . dT$

Therefore:

$$C_p . \cancel{dT} = C_v . \cancel{dT} + R . \cancel{dT}$$

$$C_p - C_v = R$$

1-7

Calculate the density of nitrogen at 75 degrees C and an absolute pressure of 700 KPa.

****

$$\rho = \frac{P}{RT} = \frac{700 \text{ KPa}}{296.5 \frac{J}{Kg \cdot °K} (75+273)°K} \times \frac{1 \text{ KN/m}^2}{\text{KPa}} \times \frac{1 J}{N \cdot m} \times \frac{10^3 N}{KN}$$

$$\rho = 6.78 \frac{Kg}{m^3}$$

1-8

Determine the pressure of carbon monoxide in a tank at a temperature of 250 °F, if its density is 0.02 slug/ft^3. Assume the gas to be under ideal conditions.

****

Absolute temperature of Carbon monoxide =

$$250 + 459.670 = 709.670 \ ^\circ R$$

Gas constant $= R = 1775.68 \ \dfrac{\text{ft-lbf}}{\text{slug-}^\circ R}$

Perfect gas law ; $p = \rho R T$

or, $p = 0.02 \times 1775.68 \times 709.670 \times \dfrac{1}{144}$

or, $p = 175.020$ psi absolute.

1-9

Determine the mass of air in a room which measures 8 x 10 x 7 ft. Assume standard atmospheric conditions (14.7 psia and 78°F). If this room is completely sealed and its temperature is increased to 85°F, determine the final pressure.

****

Volume of this room is : $V = 8 \times 10 \times 7 = 560 \text{ ft}^3$

Mass of the air in the room is : $m = \dfrac{PV}{RT} = \dfrac{14.7 \times (144) \times 560}{\dfrac{1545.3}{28.97} \times (460 + 78)}$

$$m = 41.30 \text{ lb}_m = 1.284 \text{ slugs}$$

$P_1 V_1 = m R T_1$, $P_2 V_2 = m R T_2$ can be written for two states.

For a constant volume process : $V_1 = V_2$

Therefore, $\dfrac{P_2}{P_1} = \dfrac{T_2}{T_1}$

$$P_2 = P_1 \frac{T_2}{T_1} = 14.7 \times \frac{460+85}{460+78} = 14.89 \text{ psia}.$$

SURFACE TENSION

1-10

A glass tube of 3/32 inches internal diameter is inserted into a water reservoir as shown in the figure below. The water rises in the tube as a result of surface tension effects.

Calculate this height above the level of the water in the reservoir.

Note: surface tension σ is 0.005 lb_f/ft for water-air surface at standard atmospheric conditions.

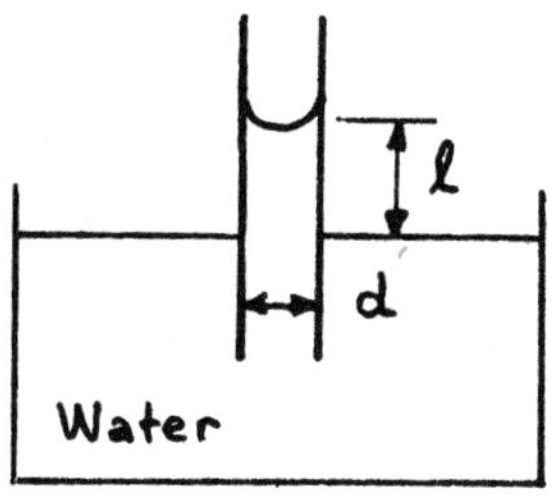

****

If we write the balance of forces in y direction:

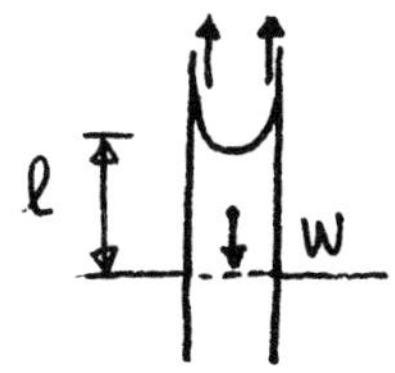

$$\Sigma F_y = 0 \quad \uparrow \overset{(+)}{y}$$

(surface tension force) − (weight of the water column) = 0

$$\delta . (\pi d) - \gamma . l . (\frac{\pi d^2}{4}) = 0$$

we get :

$$l = \frac{4\delta}{\gamma . d} = \frac{4 \times 0.005}{62.3 \times 0.0078} = 0.041 \text{ ft}$$

1-11

A capillary glass tube of 0.4 mm inside diameter is dipped in a beaker containing water at a temperature of 50°C. Determine the height of water in the tube if it stands in a vertical position.

****

From the tables of Fluid Properties at 50°C, density of water, $\rho = 988.07\ kg/m^3$, and Surface tension, $\sigma = 0.0682\ N/m$.

Capillary rise $= h = \dfrac{2\sigma \cos\theta}{\rho g r}$

or, $h = \dfrac{2 \times 0.0682 \times 1.0}{988.07 \times 9.810 \times 0.0002} = 0.070\ m$

or, $h = 70\ mm$

2 FLUID STATICS

ABSOLUTE AND GAGE PRESSURE

2-1

The pressure inside a tank is measured as 10.5 psig when the atmospheric pressure is known to be 14.4 psia. The tank is carried in the unpressurized cargo compartment of an airplane. The pressure gage in the airplane indicates that the atmospheric pressure is 21.2 inches of mercury. What does the gage on the tank read?

****

CALCULATE ABSOLUTE PRESSURE IN TANK

$$P_{ABS} = P_{GAGE} + P_{ATMOSPHERIC} = 10.5 \text{ psig} + 14.4 \text{ psia} = \underline{\underline{24.9 \text{ psia}}}$$

CALCULATE NEW ATMOSPHERIC PRESSURE

(density of H_2O)

$$P_{ATM.} = (S.G.)(\gamma)(h) = (13.57)(62.4 \text{ LB/ft}^3)\left(\frac{21.2}{12} \text{ ft. MERCURY}\right)$$

$$= 1496 \text{ psfa} = \frac{1496}{144} \text{ psia} = \underline{\underline{10.39 \text{ psia}}}$$

SOLVE FOR NEW GAGE READING

$$P_{GAGE} = P_{ABS} - P_{ATM.} = 24.9 - 10.39 = \underline{\underline{14.51 \text{ psig}}}$$

2-2

The pressure in a fluid is 100 psig and the barometer reading is 30.4 inches of mercury. Compute the absolute pressure in the fluid. The specific gravity of mercury is 13.6.

****

$$P_{absolute} = P_{gage} + P_{atmospheric}$$

$$P_{atmospheric} = (\text{Barometer Reading})(\gamma_{mercury})$$

$$\gamma_{mercury} = (sg_{mercury})(\gamma_{water})$$

$$= (13.6)(62.4 \tfrac{lb}{ft^3})$$

$$= (848.6\ lb/ft^3)\left(\frac{1 ft^3}{(12\ in)^3}\right)$$

$$\gamma_{mercury} = 0.491\ lb/in^3$$

$$P_{atmospheric} = (30.4\ in)(0.491 \tfrac{lb}{in^3})$$

$$P_{atmospheric} = 14.93\ psia$$

$$P_{absolute} = 100\ psig + 14.93\ psia$$

$$\underline{\underline{P_{absolute} = 114.93\ psia}}$$

2-3

Atmospheric pressure is 14.7 psia, the gage is reading 5.0 psi, the vapor pressure of the alcohol is 1.7 psia. Determine x and y.

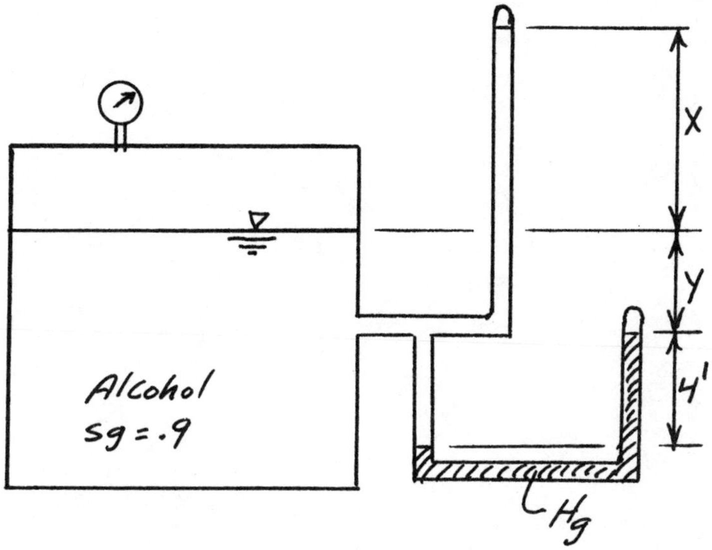

****

USE ABSOLUTE PRESSURES.

ALCOHOL MONOMETER:

$$14.7 + 5.0 - X(.9)\left(62.4 \frac{Lb}{ft^3}\right)\left(\frac{ft^2}{144\ in^2}\right) = 1.7$$

$$\underline{\underline{X = 46.1' \quad ANS}}$$

MERCURY MONOMETER.

$$14.7 + 5.0 + \frac{(Y+4)(.9)(62.4)}{144} - \frac{4(13.55)(62.4)}{144} = 0 \quad *$$

$$\underline{\underline{Y = 5.71' \quad ANS}}$$

* VAPOR PRESSURE OF MERCURY ASSUMED TO BE 0.0.

2-4

The storage tank shown below is open at the top to the atmosphere. Local atmospheric pressure is 780 mm Hg. The Bourdon gage at the bottom of the tank reads 50 psi. The temperature of the water and the glycerin is 68 degrees F. Determine the depth of the water in the tank.

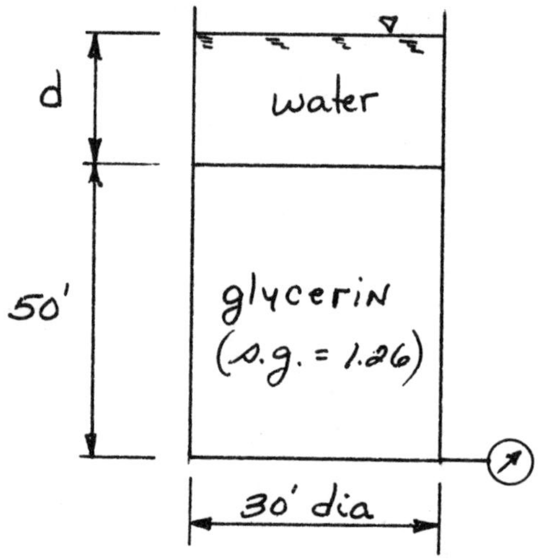

****

$$50\ psi = P_{water} + P_{glycerin} = \left[(\gamma h)_{water} + (\gamma h)_{glycerin}\right]\frac{g}{g_c}$$

$$50\ \frac{lb_f}{in^2} \times \frac{144\ in^2}{ft^2} = \left[\left(62.4\ \frac{lb_m}{ft^3}\right)(d) + \left(1.26 \times 62.4\ \frac{lb_m}{ft^3}\right)(50\ ft)\right]\frac{32\ ft/s^2}{32\ \frac{ft\ lb_m}{lb_f\ s^2}}$$

$$7200 = 62.4d + 3931.2 \Rightarrow \underline{\underline{d = 52.4\ ft}}$$

2-5

Find the absolute pressure inside container B if the barometric pressure is 720 mm of mercury.

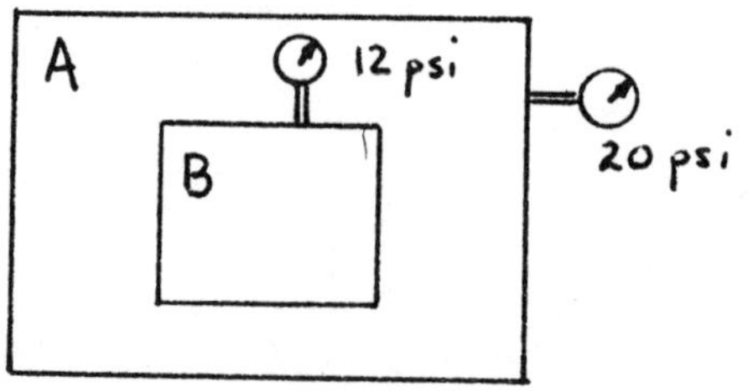

****

$$P_{bar} = \frac{720 \text{ mm}}{760 \text{ mm}} * 14.7 \text{ psi} = 13.9 \text{ psia}$$

$$P_{abs_A} = 13.9 + 20 = 33.9 \text{ psia}$$

$$P_{abs_B} = 33.9 + 12 = \underline{45.9 \text{ psia}}$$

MANOMETERS

2-6

In the figure shown below determine p_B if p_A=20 psia.

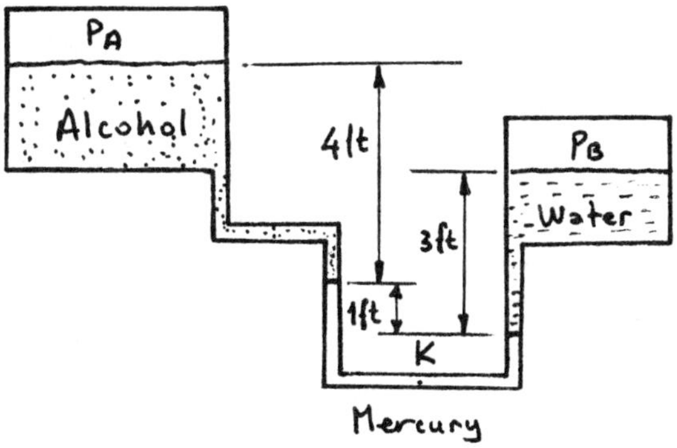

****

If we write the balance of forces at point K:

$$P_A + 4\text{ft} \cdot \gamma_a + 1\text{ft} \cdot \gamma_m = 3\text{ft} \cdot \gamma_w + P_B$$

$$20 \times 144 + 4 \times (0.8 \times 62.3) + 1 \times 846 = 3 \times 62.3 + P_B$$

$$P_B = 3738.46 \frac{\text{lbf}}{\text{ft}^2} = 25.96 \text{ psia}$$

2-7

Determine the gage pressure and absolute pressure of water in the vessel at the point A as shown. The heavier liquid in the manometer is mercury. Assume density of water, ρ_1 = 1 000 kg/m^3, density of mercury, ρ_2 = 13 550 kg/m^3, and standard atmospheric pressure p_0 = 101 325 Pa. Consider x = 50 mm and y = 200 mm in the figure shown. Use the value of gravitational acceleration, g = 9.810 m/s^2.

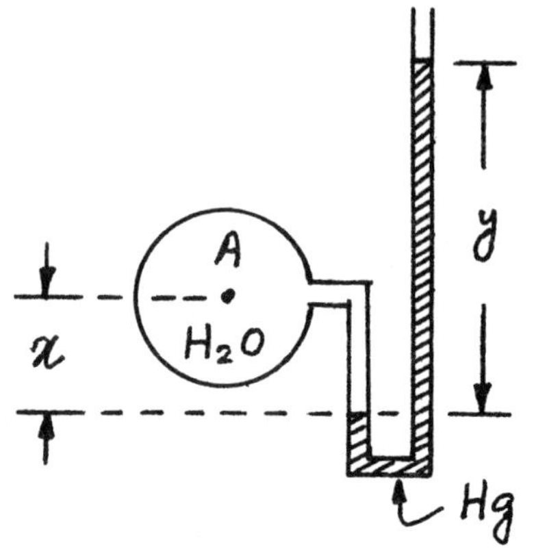

****

density

Gage pressure at A = $(p_A)_{gage} = \frac{g}{g_c}(\rho_2 y - \rho_1 x)$

$= \frac{9.810}{1}(13,550 \times 0.200 - 1,000 \times 0.050)$

pressure at point A

or, $(p_A)_{gage}$ = 26.100 kPa

Absolute Pressure at A = $(p_A)_{absolute} = (p_A)_{gage} + p_0$

or, $(p_A)_{absolute}$ = $26.100 + 101.325 \simeq 127.4$ kPa

2-8

Two pipelines containing oil (s.g. = 0.86) are connected by a differential manometer as shown below. The pressure in pipe A is 12.6 psi. Determine the pressure in pipe B. The specific gravity of mercury is 13.57.

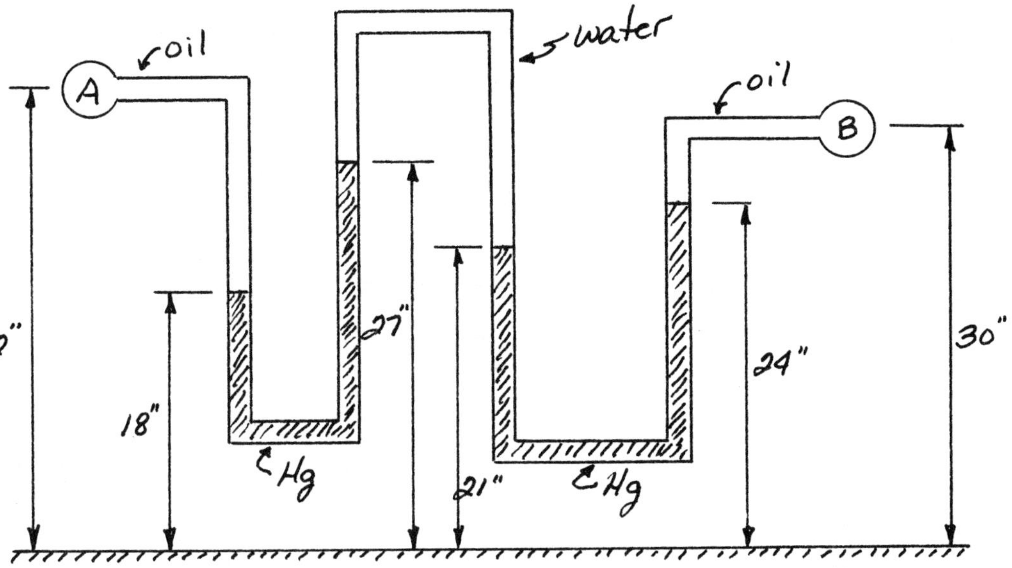

****

$$P_A + \Sigma \gamma h = P_B$$

$$\gamma_{oil} = .86 \times 62.4 \frac{lb}{ft^3} \times \frac{1\,ft^3}{1728\,IN^3} = .031 \frac{lb}{IN^3}$$

$$\gamma_{Hg} = 13.57 \times 62.4 \frac{lb}{ft^3} \times \frac{1\,ft^3}{1728\,IN^3} = .490 \frac{lb}{IN^3}$$

$$\gamma_{H_2O} = 62.4 \frac{lb}{ft^3} \times \frac{1\,ft^3}{1728\,IN^3} = .036 \frac{lb}{IN^3}$$

$$P_A + \left(.031 \frac{lb}{IN^3}\right)(14") - .490 \frac{lb}{IN^3}(9") + .036 \frac{lb}{IN^3}(6") - .490 \frac{lb}{IN^3}(3")$$

$$- .031 \frac{lb}{IN^3}(6") = P_B$$

$$\left[12.6 + .434 - 4.41 + .216 - 1.47 - .186\right] psi = P_B$$

$$\underline{\underline{P_B = 7.2\ psi}}$$

2-9

The manometer shown is open to the atmosphere at Point B. Find the gage pressure at Point A.

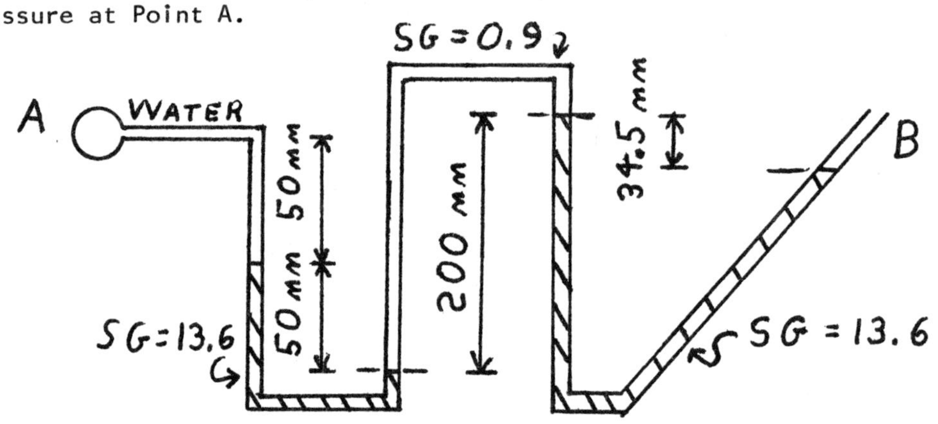

****

GAGE PRESSURE AT $B = 0\ kN/m^2$

$\gamma_{WATER} = 9.81\ kN/m^3$

$$P_A + .05m(\gamma_w) + .05m(13.6)\gamma_w - .2m(0.9)\gamma_w + .0345(13.6)\gamma_w = P_B = 0$$

$$P_A + .49\ kN/m^2 + 6.67\ kN/m^2 - 1.77\ kN/m^2 + 4.60\ kN/m^2 = 0$$

$$P_A + 9.99\ kN/m^2 = 0$$

$$P_A = -9.99\ kN/m^2 \text{ (gage)}$$

2-10

Calculate the pressure at points A and B in the manometer system shown.

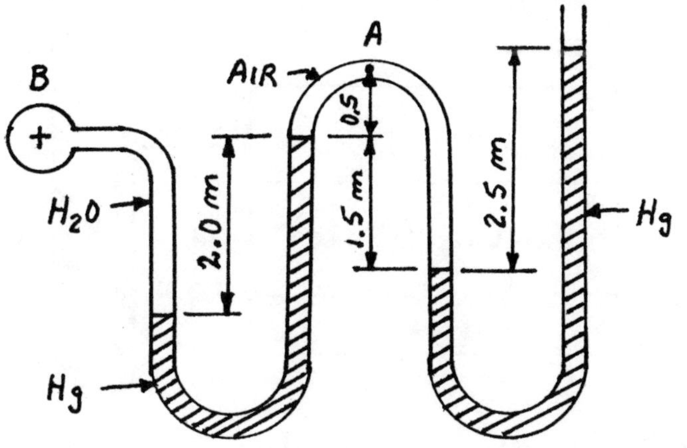

****

ASSUME $P_{ATM.}$ = 0 PSIG AND SOLVE BY STARTING AT OPEN END OF MANOMETER.

$$P_A = P_{ATM} + \gamma_{Hg} h_1 - \gamma_{AIR} h_2 \qquad \text{NEGLECT } \gamma_{AIR} \approx 0$$

$$\gamma_{Hg} = S.G.\ \rho_H\ g = (13.57)(998.2\ kg/m^3)(9.81\ m/s^2)$$

$$= 132{,}882\ \left(\frac{kg\text{-}m}{s^2}\right)\left(\frac{1}{m^3}\right) = 132.88\ kN/m^3$$

$$P_A = 0 + (132.88\ \frac{kN}{m^3})(2.5\ m) = \underline{\underline{332.2\ kPa}}$$

$$P_B = P_{ATM} + \gamma_{Hg}(2.5) - \gamma_{AIR}(1.5) + \gamma_{Hg}(2.0) - \gamma_{H_2O}(2)$$

$$\gamma_{H_2O} = \rho_{H_2O}\ g = 998.2\ (kg/m^3)(9.81\ m/s^2) = 9{,}792\left(\frac{kg\text{-}m}{s^2}\right)\frac{1}{m^3}$$

$$\gamma_{H_2O} = 9.792\ kN/m^3$$

$$P_B = 0 + (132.88)(2.5) - (0) + 132.88(2.0) - 9.792(2)$$

$$P_B = \underline{\underline{578.4\ kPa}}$$

2-11

What is the manometer reading (R) if the gage reads 200 kPa?

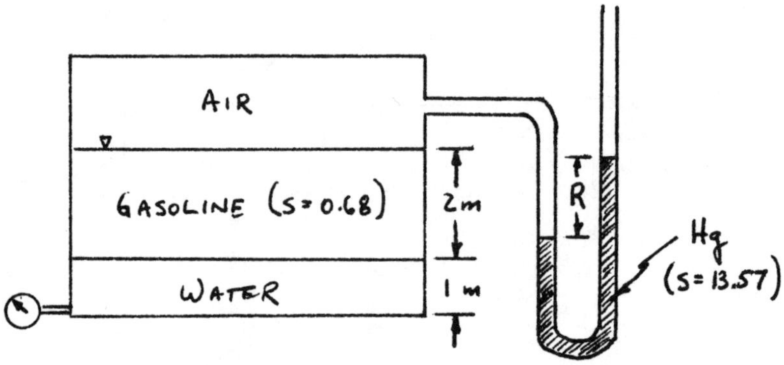

****

Write manometer equation (in m. H_2O) from top of mercury to pressure gage.

$$13.57\,R + 2\text{m}\,(0.68) + 1\text{m}\,(1.0)$$

$$= 200\text{ kPa} \times \left(\frac{10.3\text{ m}}{101.3\text{ kPa}}\right) \leftarrow \text{unit conversion}$$

$$13.57\,R = 20.34 - 1.36 - 1.00$$

$$\underline{R = 1.32\text{ m}}$$

2-12

The Manometer shown is open to the atmosphere. Determine the gage reading, P.

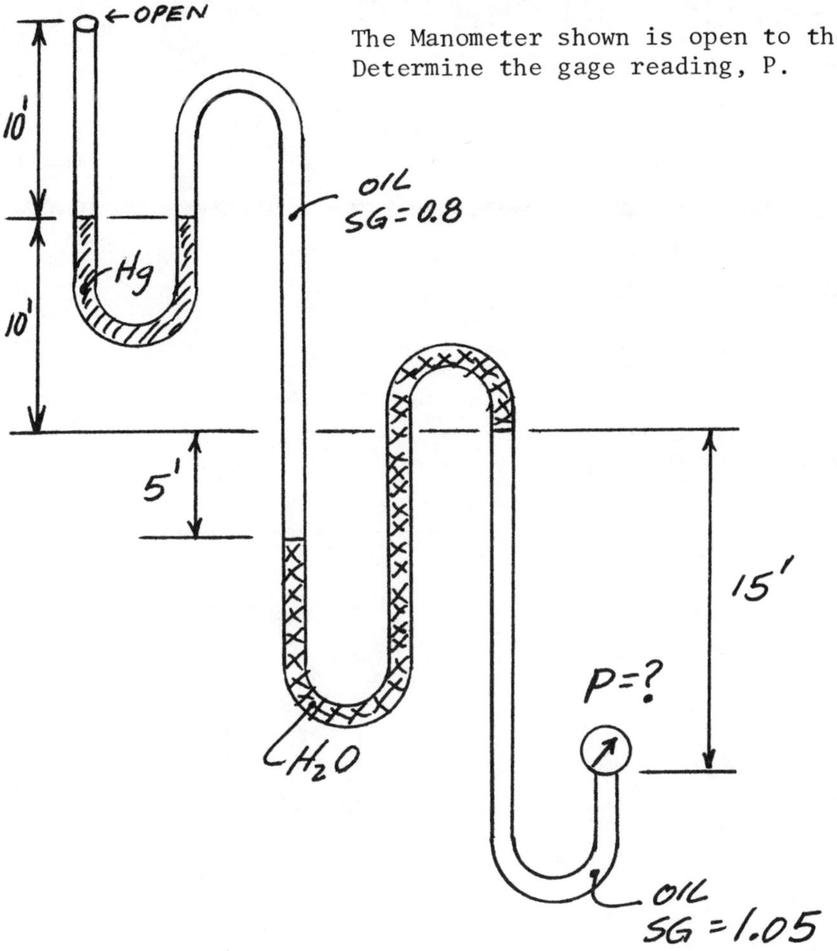

****

$$\frac{0 + 15(.8)(62.4) - 5(62.4) + 15(1.05)(62.4)}{144} = P$$

$P = 9.86$ psi ANS

FORCES ON SURFACES

2-13

Due to gage pressure only, determine the hydrostatic force and its point of application on one side of the triangle as shown. The triangle is dipped in water with density of 1,000 kg/m^3. Assume gravitational acceleration, g = 9.810 m/s^2. Measurements are exact.

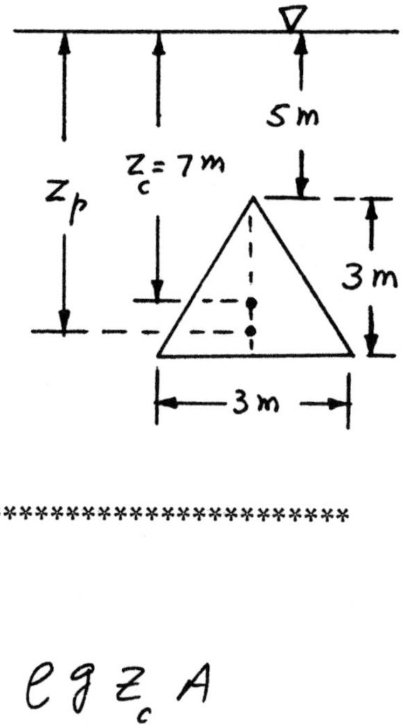

****

The hydrostatic force = $F = \rho g z_c A$

or, $F = \frac{1000}{1000} \times 9.810 \times 7 \times \left(\frac{3 \times 3}{2}\right) \simeq 309.0 \text{ kN}$

Point of application of force $F = z_p = z_c + \frac{I_c}{z_c A}$

Moment of inertia of triangle $= I_c = \frac{3 \times (3)^3}{36}$

Then, $z_p = 7 + \frac{3 \times (3)^3/36}{7 \times 4.5} = 7.071 \text{ m}$

2-14

A 5' x 7' rectangular steel gate is used as a plug in the water storage system shown below. A motor-driven winch is used to open the gate by a cable connected to the end of the gate. Determine the tension in the cable in order to open the gate. Neglect the weight of the gate.

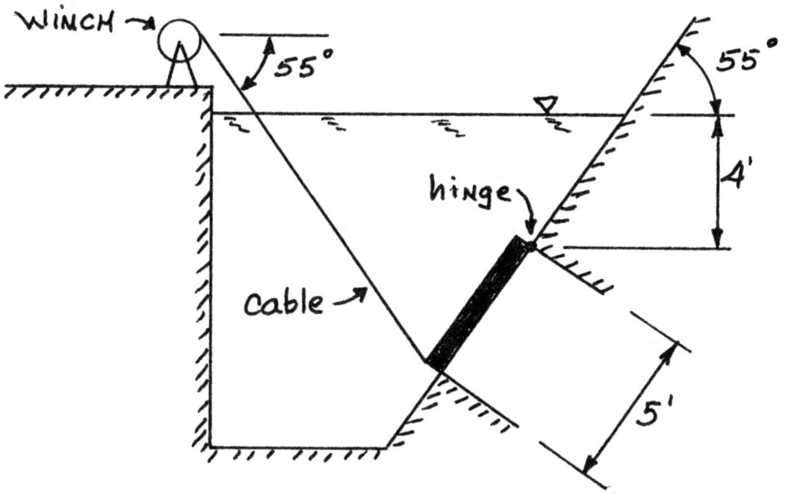

****

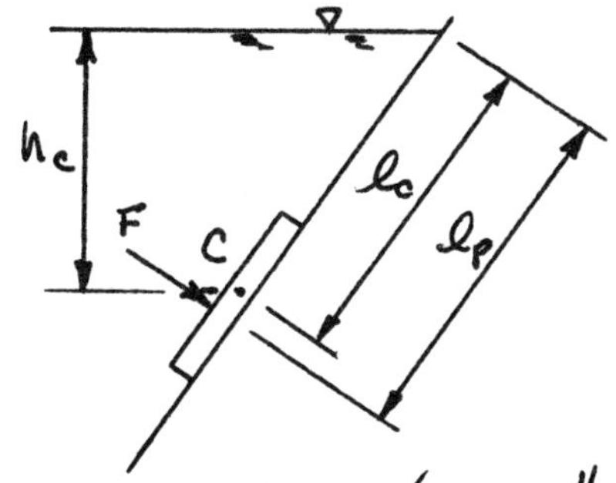

$$h_c = 4' + 2.5' \sin 55° = 6.05'$$

$$l_c = \frac{6.05'}{\sin 55°} = 7.38'$$

$$I_c = \frac{bh^3}{12} = \frac{7'(5')^3}{12} = 72.92 \text{ ft}^4$$

$$F = \gamma h_c A = \left(62.4 \frac{\text{lb}}{\text{ft}^3}\right)(6.05')(7' \times 5') = 13213.2 \text{ lb}$$

$$l_p = l_c + \frac{I_c}{l_c A} = 7.38' + \frac{72.92 \text{ ft}^4}{(7.38')(35 \text{ ft}^2)} = 7.66'$$

<u>FBD of gate</u>: (assume frictionless hinge).

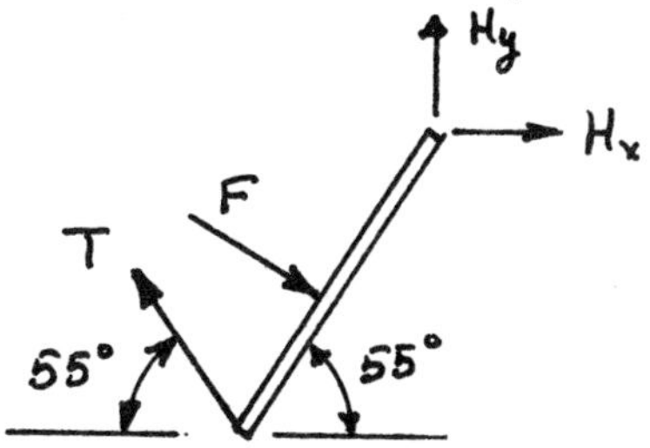

$$\overset{+}{\curvearrowright} \sum M_{Hinge} = 0$$

$$(T \sin 70)(5') - (13213.2 \text{ lb})\left(7.66' - \frac{4'}{\sin 55}\right) = 0$$

$$\underline{\underline{T = 7809 \text{ lb}}}$$

2-15

The bottomless tank shown in the figure below has a weight of 1000 lb .It is filled with water at 70°F. Six bolts on the right and left hand sides hold the tank down. Determine the force acting on each bolt.

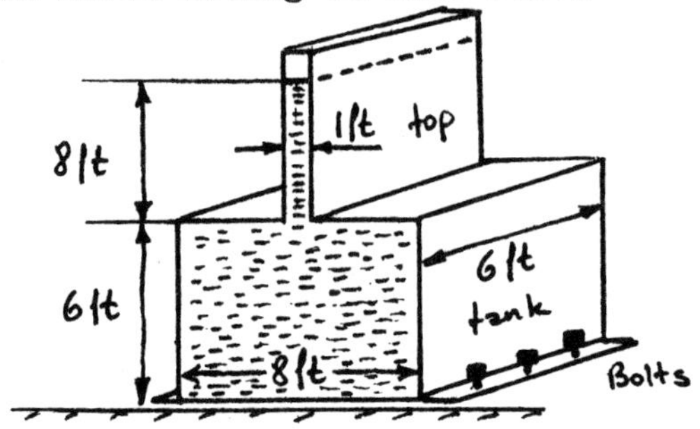

****

If we write the balance of forces in the y direction, we get

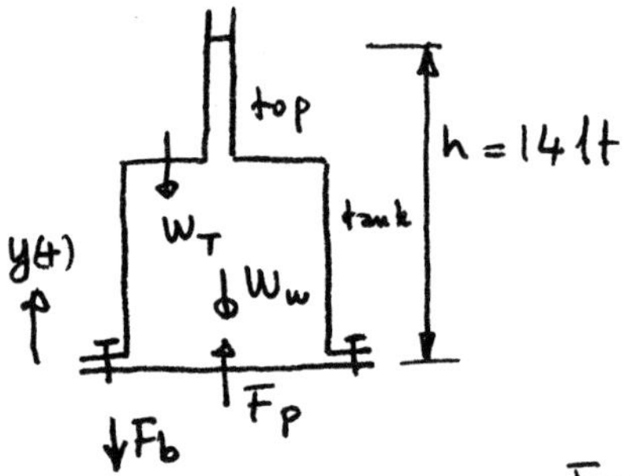

$$\Sigma F_y = 0$$

$$-F_b + F_p - W_T - W_w = 0 \quad \text{which gives}$$

$$F_b = F_p - W_T - W_w$$

$$F_p = h \cdot \gamma_w \cdot A_{bottom} = 14 \times 62.4 \times 48 = 41932.8 \text{ lbf.}$$

$$W_w = (W_w)_{top} + (W_w)_{tank} = \gamma \left(V_{top} + V_{tank} \right)$$

$$= 62.4 \times \left(1 \times 8 \times 6 + 8 \times 6 \times 6 \right) = 20966.4 \text{ lbf}$$

$$F_b = 41932.8 - 1000 - 20966 = 19966.4 \text{ lbf}$$

For each bolt $\quad F = F_b / 6 = 3327.73 \text{ lbf}$.

2-16

A pair of gates at one end of a river lock retains water within the enclosure. Find the torque which acts to open one gate(i.e., the y-component of torque on the gate). Hint: What is the force on a small element of area? What is its contribution to the torque about the hinge? Consider the contributions for all such area elements.

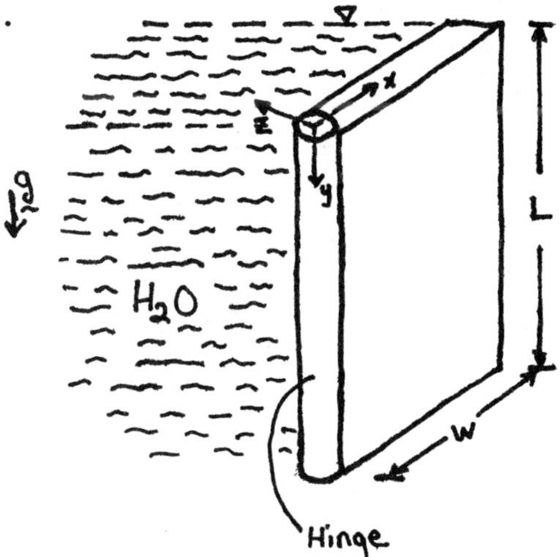

W = 8 ft.

L = 15 ft.

****

For an infinitesimal area element dA which is located at distance x from the hinge and at depth y, the moment arm is $\underset{\sim}{r} = x\underset{\sim}{i}$ and the force by the hydrostatic pressure is $d\underset{\sim}{F} = -p\,dA\,\underset{\sim}{k}$ (Note that dF increases with y). Contribution of this force to the torque is:

$$d\underset{\sim}{M} = \underset{\sim}{r} \times d\underset{\sim}{F} \quad \Rightarrow \quad \underset{\sim}{M} = \int \underset{\sim}{r} \times d\underset{\sim}{F}$$

The torque about the hinge is the y-component.

$$M_{Hinge}\,\underset{\sim}{j} = \int x\underset{\sim}{i} \times (-p\,dA)\underset{\sim}{k} \quad \text{or} \quad M_{Hinge} = \int_A x p\,dA = \int_0^L\int_0^W x p\,dx\,dy$$

Pressure must be expressed as a function of depth y before integrating.

$$M_{Hinge} = \int_0^L\int_0^W x(\rho_{H_2O}\, g\, y)\,dx\,dy = \rho_{H_2O}\, g \int_0^L\int_0^W xy\,dx\,dy = \frac{\rho_{H_2O}\, g\, W^2}{2}\int_0^L y\,dy$$

$$M_{Hinge} = \frac{\rho_{H_2O}\, g\, W^2 L^2}{4} = \frac{(62.4\ lb_f/ft^3)(8ft)^2(15ft)^2}{4} = 2.25\cdot 10^5\ ft\text{-}lb_f$$

2-17

A rectangular area 2'x4' lies below the water surface as shown.

a) Determine the magnitude, direction and location of the resultant force on the area.

b) Also determine the magnitude by the method of horizontal and vertical components.

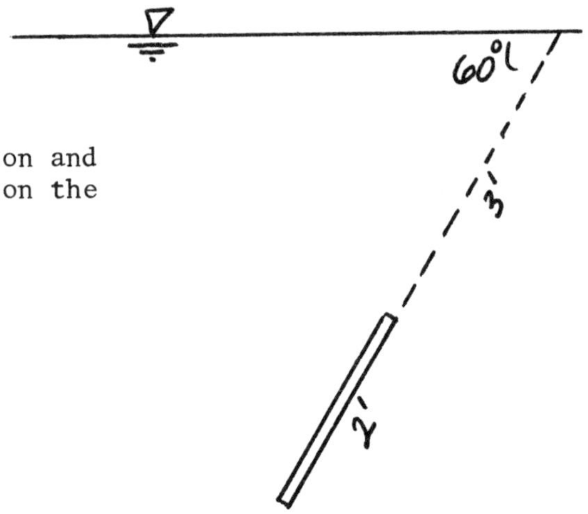

****

a)

$$F = \gamma d_c A$$

$$F = \left(62.4 \frac{Lb}{ft^3}\right)\left(4 \sin(60)\right)\left(8 \, FT^2\right) \qquad d_c = 4 \sin(60)$$

$= 1729^{\#}$ (30°) ANS

$$\Delta L = \frac{I_c}{L_c A} \; ; \; I = \frac{bh^3}{12}$$

$$= \frac{\frac{4(2)^3}{12}}{(4)(8)} = .0833' \text{ below C.G.}$$ ANS.

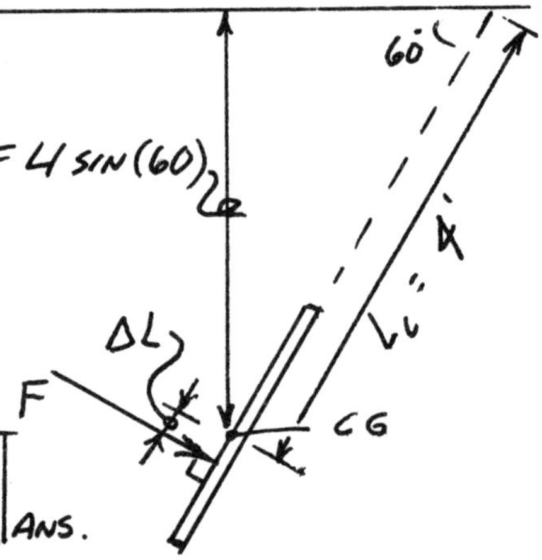

b) $W_1 = 2.60(1)(4)(62.4) = 649^{\#}$

$W_2 = \frac{1}{2}(1)(1.73)(4)(62.4) = 216^{\#}$

$F_1 = \gamma h_c A = 62.4(4 \sin 60)(1.73 \times 4) = 1496^{\#}$

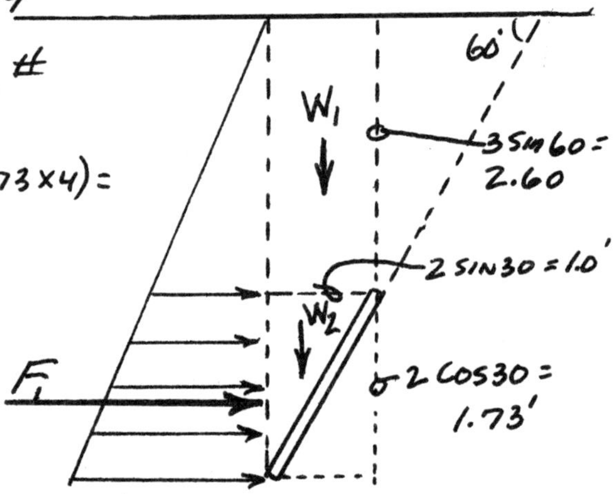

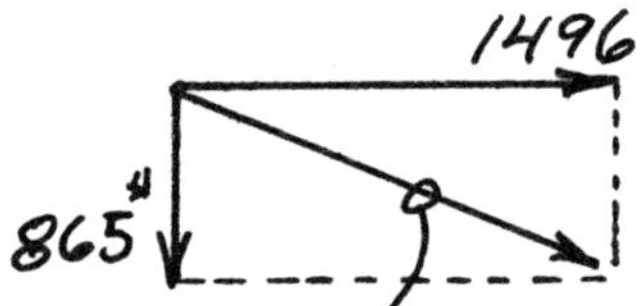

$\sqrt{(1496)^2 + (865)^2} = 1728^{\#}$ ANS

2-18

Calculate the magnitude and location of the force on the submerged circular gate. What moment must be applied at the supports to keep the gate from opening?

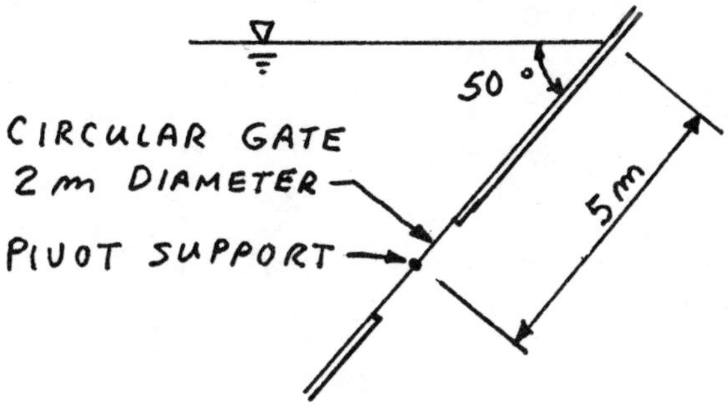

****

SOLVE FOR FORCE, F.

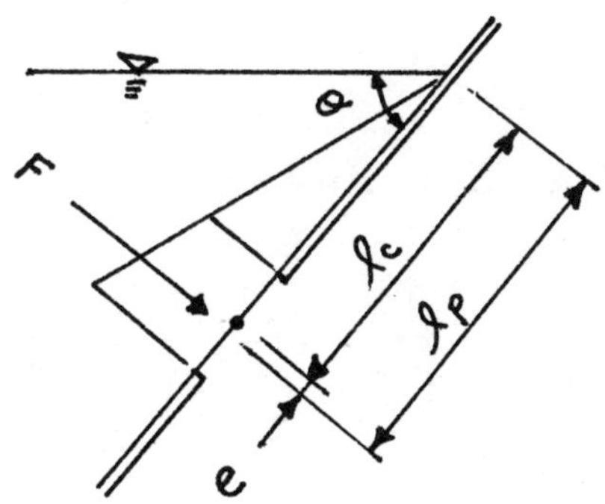

$$F = \gamma h_c A = \gamma l_c \sin\theta\, A$$

$$l_c = 5\text{ m}$$

$$\gamma_{H_2O} = \rho g = (998.2\text{ Kg/m}^3)(9.81\text{ m/s}^2)$$

$$\gamma_{H_2O} = 9.792\text{ kN/m}^3$$

$$A = \frac{\pi D^2}{4} = 3.1416\frac{(2)^2}{4} = 3.14\text{ m}^2$$

$$F = \left(9.792\frac{\text{kN}}{\text{m}^3}\right)(5\text{ m})(\sin 50^\circ)(3.14\text{ m}^2) = \underline{\underline{117.8\text{ kN}}}$$

FIND LOCATION OF FORCE, F.

$$l_p = l_c + I_c/l_c A \qquad \text{WHERE } I_c = \frac{\pi D^4}{64} = \frac{\pi(2)^4}{64} = 0.785\text{ m}^4$$

$$l_p = 5\text{ m} + \frac{0.785\text{ m}^4}{(5\text{ m})(3.14\text{ m}^2)} = \underline{\underline{5.05\text{ m}}} \qquad \text{AND } e = l_p - l_c = \underline{\underline{0.05\text{ m}}}$$

CALCULATE RESISTING MOMENT

$$M = Fe = (117.8\text{ kN})(0.05\text{ m}) = \underline{\underline{5.89\text{ kN-m}}}$$

2-19

A tank conatins a liquid whose pressure-density relation is $\rho = kp^{2/3}$ where k is a constant. A door of length L and width b is in the tank a distance h below the surface, as shown.

a) Determine the force from the fluid acting on the door.
b) Determine the center of pressure of the door.

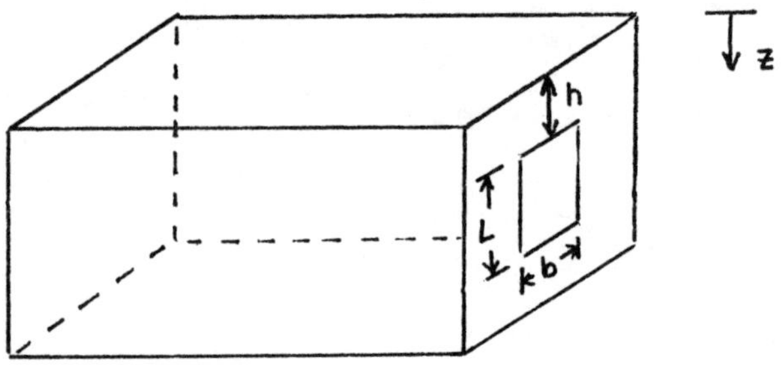

****

a) LET z BE THE VERTICAL COORDINATE, POSITIVE DOWNWARD. THE DIFFERENTIAL RELATION BETWEEN PRESSURE AND z IS

$$\frac{dp}{dz} = +\gamma = +\rho g = +gKp^{2/3} \qquad (1)$$

AT $z=0$ $p=0$

SEPARATING THE VARIABLES IN EQUATION (1)

$$\frac{dp}{p^{2/3}} = gK\,dz$$

INTEGRATING BOTH SIDES

$$3p^{1/3} = gKz + C \quad , \quad C=0 \text{ FROM BOUNDARY CONDITION}$$

$$p = \left(\frac{gK}{3}z\right)^3$$

THE TOTAL FORCE ON THE DOOR IS

$$F = \int p\,dA \qquad dA = b\,dz$$

$$F = \int_h^{h+L} b\left(\frac{gK}{3}z\right)^3 dz = \frac{b}{4}\left(\frac{gK}{3}\right)^3 z^4 \Big|_h^{h+L}$$

$$F = \frac{b}{4}\left(\frac{gK}{3}\right)^3\left[(h+L)^4 - h^4\right]$$

b) THE CENTER OF PRESSURE, BY DEFINITION, IS THE POINT WHERE THE RESULTANT OF THE PRESSURE FORCES SHOULD BE PLACED. THUS THE MOMENT OF THE RESULTANT FORCE PLACED AT THE CENTER OF PRESSURE ABOUT ANY POINT SHOULD EQUAL THE TOTAL MOMENT OF THE PRESSURE DISTRIBUTION ABOUT THIS POINT. LET $z_{c.p.}$ BE THE VERTICAL DISTANCE FROM THE SURFACE TO THE CENTER OF PRESSURE, THEN

$$F z_{c.p.} = \int_h^{h+L} z\, p(b\, dz)$$

$$F z_{c.p.} = \int_h^{h+L} b\left(\frac{gK}{3}\right)^3 z^4\, dz$$

$$F z_{c.p.} = \frac{b}{5}\left(\frac{gK}{3}\right)^3 z^5 \Big|_h^{h+L}$$

$$F z_{c.p.} = \frac{b}{5}\left(\frac{gK}{3}\right)^3 \left[(h+L)^5 - h^5\right]$$

OR

$$\frac{b}{4}\left(\frac{gK}{3}\right)^3 \left[(h+L)^4 - h^4\right] z_{c.p.} = \frac{b}{5}\left(\frac{gK}{3}\right)^3 \left[(h+L)^5 - h^5\right]$$

$$z_{c.p.} = \frac{4}{5}\,\frac{(h+L)^4 - h^4}{(h+L)^5 - h^5}$$

2-20

A floating cylinder is attached to the top of a dam to increase its height. Assuming that the gate is in equilibrium in the position shown, determine the amount of water ballast that is contained within the cylinder. The cylinder has a weight of 150 lb/LF.

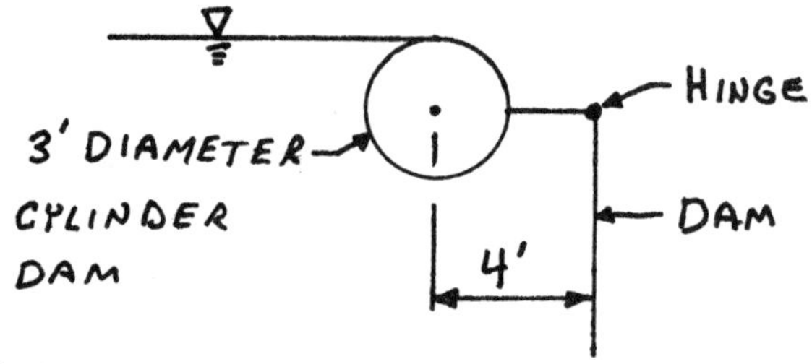

****

CALCULATE BUOYANT FORCES FOR 1' DEPTH INTO PAPER.

$$F_{B_1} = \gamma_{H_2O} \forall_{DISPLACED} = \gamma \frac{\pi}{2} \frac{D^2}{4}$$

$$= \left(62.4 \frac{LB}{Ft^3}\right) \frac{1}{2} \pi \frac{(3)^2}{4}$$

$$F_{B_1} = \underline{\underline{220.5 \text{ LB} \uparrow}}$$

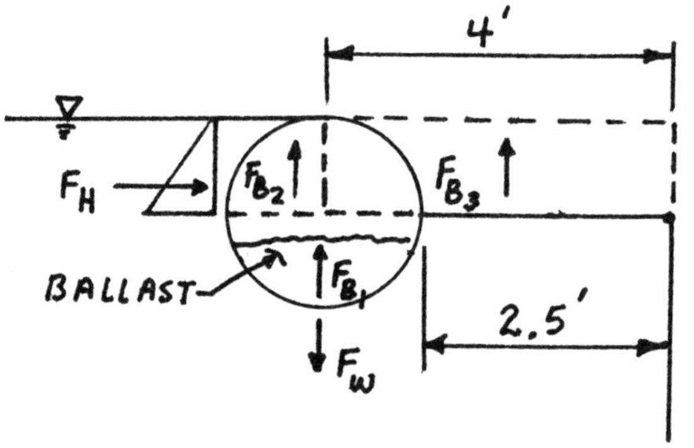

$$F_{B_2} = \gamma_{H_2O} \forall_D = \gamma \frac{\pi}{4} \frac{D^2}{4} = \left(62.4 \frac{LB}{Ft^3}\right) \frac{1}{4} \pi \frac{(3)^2}{4} = \underline{\underline{110.3 \text{ LB} \uparrow}}$$

$$F_{B_3} = \gamma_{H_2O} \forall_D = \left(62.4 \frac{LB}{Ft^3}\right)(1.5 \text{ Ft.} \times 4.0 \text{ Ft.}) = \underline{\underline{374.4 \text{ LB} \uparrow}}$$

SOLVE FOR HORIZONTAL FORCE RESULTANT. FORCES ON BOTH SIDES OF LOWER HALF OF CYLINDER CANCEL EACH OTHER.

$$F_H = \gamma h_c A = \gamma \left(\frac{D}{2}\right)\left(\frac{D}{2} \times 1\right) = \gamma \frac{D^2}{4} = (62.4) \frac{(3)^2}{(4)} = \underline{\underline{140.4 \text{ LB} \rightarrow}}$$

SOLVE FOR MOMENT ARMS ABOUT HINGE AT O

$$x = 4 + \frac{4R}{3\pi} = 4 + \left(\frac{4}{3}\right)\left(\frac{1}{\pi}\right)\left(\frac{D}{2}\right) = 4.64'$$

$$y = \frac{h}{3} = \frac{R}{3} = \frac{D/2}{3} = \frac{3/2}{3} = 0.5'$$

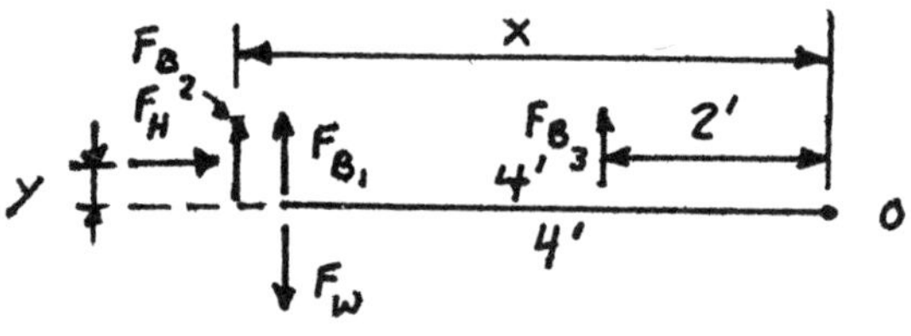

SOLVE FOR F_W (BALLAST) BY TAKING MOMENTS ABOUT O

$$\overset{\curvearrowright +}{\Sigma M_o} = 0 = F_{B_1}(4) + F_{B_2}(4.46) + F_{B_3}(2) + F_H(0.5) - F_W(4)$$

$$(220.5)(4) + (110.3)(4.46) + 374.4(2) + 140.4(0.5) = F_W(4)$$

$$F_W = 548.2 \text{ LB/LF}$$

WATER BALLAST = F_W - CYLINDER WEIGHT = 548.2 - 150 = 398.2 LB/LF

CHECK MAXIMUM BALLAST CAPACITY = $\gamma V_D = (62.4)(\pi \frac{D^2}{4})$

$$= (64.2 \text{ LB/FT}^3)(\pi)\frac{(3)^2}{(4)} = 441 \text{ LB.}$$ BALLAST OKAY

2-21

The tank shown is 2.0 m in length. Compute the resultant force, F_R, on side BC and the distance X to the center of pressure, cp.

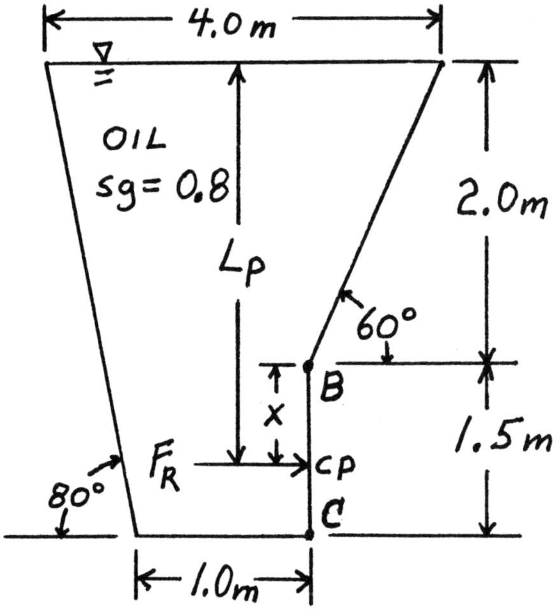

****

$F_R = \gamma\, d_c\, (AREA)_{BC}$

d_c = depth to centroid of side BC

$d_c = 2.0\,m + 1.5\,m/2$

$d_c = 2.75\,m$

$AREA_{BC} = (1.5\,m)(2.0\,m)$

$AREA_{BC} = 3\,m^2$

$F_R = (0.8)(9.81\,\frac{kN}{m^3})(2.75\,m)(3.0\,m^2)$

$F_R = 64.75\,kN$

$L_p = L_c + I/L_c A$

$L_c = d_c = 2.75\,m$

$$I = \frac{(base)(height)^3}{12}$$

$I = (2\,m)(1.5\,m)^3/12$

$I = 0.56\,m^4$

$$L_p = 2.75\,m + \frac{0.56\,m^4}{(2.75\,m)(3\,m^2)}$$

$L_p = 2.75\,m + 0.07\,m$

$L_p = 2.82\,m$

$X = L_p - 2.0\,m$

$X = 2.82\,m - 2.0\,m$

$\underline{X = 0.82\,m}$

BUOYANCY AND STABILITY

2-22

A 300 foot long, 25 foot diameter nuclear submarine has sunk off the coast of New Jersey. The sub weighs 5000 tons empty, but carries an additional 750 tons of ballast. Because of the danger to our national defenses, the submarine must be returned to the surface and towed back ashore. In order to float the sub, how many tons of ballast must be pumped out of the vessel? Assume the specific gravity of seawater to be 1.2

****

$$\forall = \frac{\pi (25)^2}{4}(300) = 1.47 * 10^5 \ ft^3$$

Buoyant Force, $F_b = 1.2\left(62.4 \frac{lb}{ft^3}\right)(1.47 * 10^5 \ ft^3)$

$$F_b = 1.1 * 10^7 \ lbs = 11 \ M \ lbs.$$

$$F_b = W_{sub} + W_{ball.}$$

$$W_{sub} = 5000 * 2000 = 10 \ M \ lbs.$$

$$\therefore \ 11 \ M \ lbs = 10 \ M \ lbs + W_{ball}$$

$$\Rightarrow W_{ballast} = 1 \ M \ lbs. \text{ needed to float.}$$

Initially $W_{ball} = 750 * 2000 = 1.5 \ M \ lbs.$

Ballast removed $= (1.5 - 1.0) = 0.5 \ M \ lbs$

$$= 250 \text{ tons}$$

2-23

An ice cube is floating in a brimfull cup of water as shown.

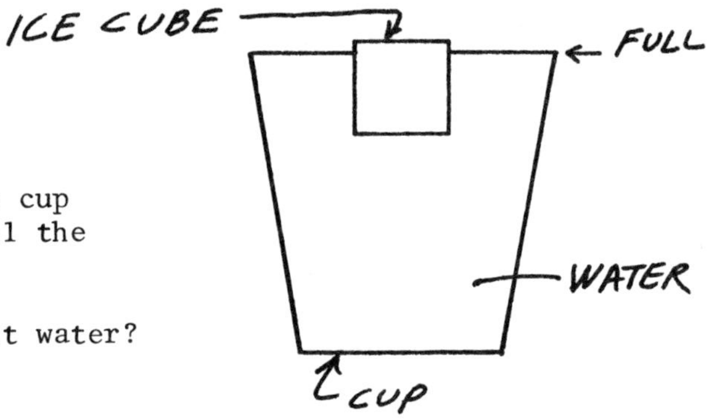

a) As the ice cube melts will the cup overflow, stay the same or will the level drop?

b) What if the cup is full of salt water?

****

a) stay the same. The wt. of the ice cube equals the wt. of the water displaced. (i.e. same quantity of H_2O)

b) Overflow. A smaller volume of salt water is displaced since it is more dense than fresh water.

2-24

A person holding 1 ft^3 of concrete (150 lb/ft^3) is in a raft floating in a pool.

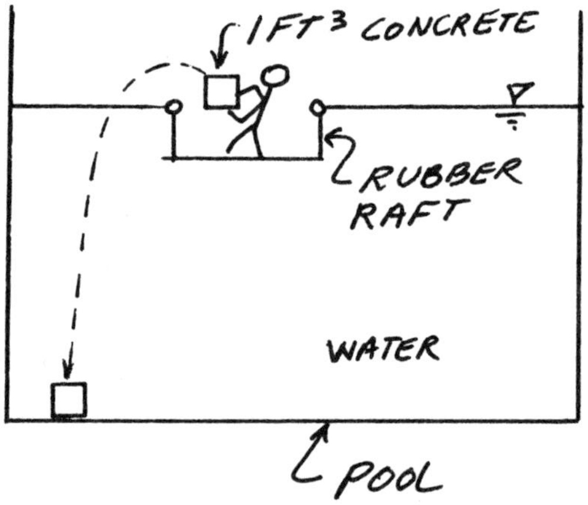

The concrete is tossed overboard. Will the water level in the pool rise, lower or stay the same?

****

When the Concrete is floating, its Weight must be displaced. This would be about $\frac{150}{62.4} = 2.40$ ft³ of water.

When the Concrete sinks, only its Volume (1 ft³) will be displaced.

Therefore the level will lower.

2-25

A sample of course gravel was found to weigh 37.5 lb. in the atmosphere. The sample was completely submerged in water and weighed again. If the weight when submerged is 6.3 lb., what is the specific weight of the gravel?

****

DRAW A FREEBODY DIAGRAM FOR THE SUBMERGED WEIGHT

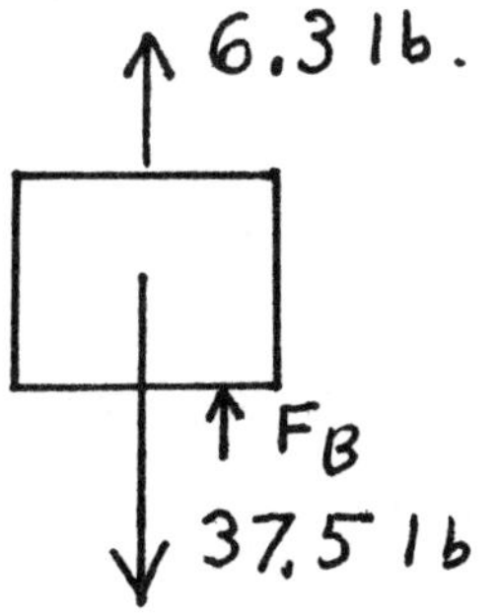

$$\Sigma F_{VERTICAL} = 0$$

$$6.3 + F_B = 37.5$$

$$F_B = 31.2 \text{ lb}$$

F_B = Buoyant Force
$= \gamma \cdot$ Displaced Volume
$= \gamma_w \cdot V_d$
$= 62.4 \text{ lb/Ft}^3 \cdot V_d$

So $31.2 \text{ lb} = 62.4 \text{ lb/Ft}^3 \cdot V_d$

$$V_d = \frac{31.2}{62.4} = 0.5 \text{ Ft}^3$$

Specific Weight of Gravel $= \frac{37.5 \text{ lb}}{0.5 \text{ Ft}^3}$

$$\gamma_G = 75 \frac{\text{lb}}{\text{Ft}^3}$$

2-26

How large a cable force must be supplied to support a 3 ft. cube which weighs 2000 pounds in air? The oil has a specific gravity of 0.8 and water has a density of 1.94 slugs/ft^3.

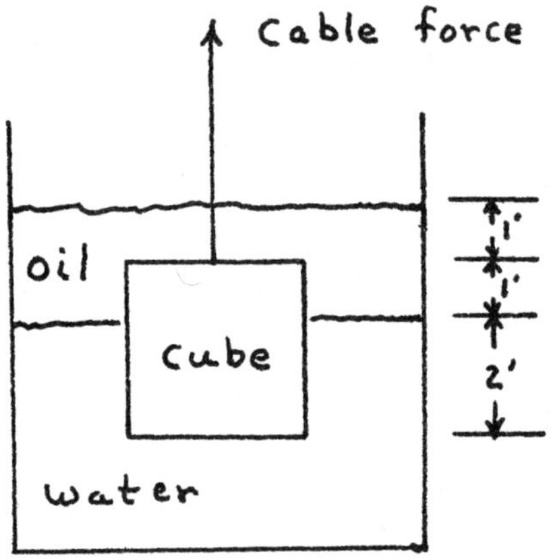

****

Pressure on upper surface of cube. $(\gamma = \rho g)$ $\left(\text{slug} = \frac{\text{lb}_f\text{-sec}^2}{\text{ft}}\right)$

$$p_{up} = \rho_{oil}\, g h = .8(1.94)\ \frac{\text{lb}_f\ \text{sec}^2}{\text{ft}^4}\ 32.2\ \frac{\text{ft}}{\text{sec}^2}\ 1\ \text{ft} = 50.0\ \frac{\text{lb}_f}{\text{ft}^2}$$

pressure on lower surface of cube,

$$p_{low} = p_{up} + \rho_{oil}\, g(1) + \rho_{water}\, g(2)$$

$$p_{low} = 50.0 + .8(1.94)(32.2)(1) + 1.94(32.2)(2) = 224.9\ \frac{\text{lb}_f}{\text{ft}^2}$$

Buoyant force is due to the 2 pressure forces

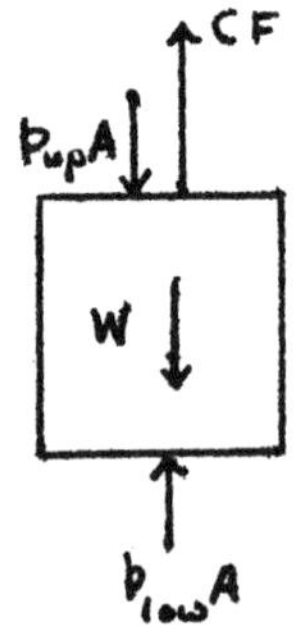

$$B.F. = p_{low}A - p_{up}A = 224.9(3\times3) - 50.0(3\times3)$$

$$B.F. = 2024 - 450 = 1574\ \text{lb}_f$$

From free body,

$$\text{Cable force} = CF = W - BF = 2000 - 1574 = 426\ \text{lb}_f$$

2-27

A large rock (s.g. = 2.15) is submerged in water and suspended by a wire attached to the end of a rigid beam as shown below. The system is in equilibrium when the beam is horizontal. Determine the weight of the rock.

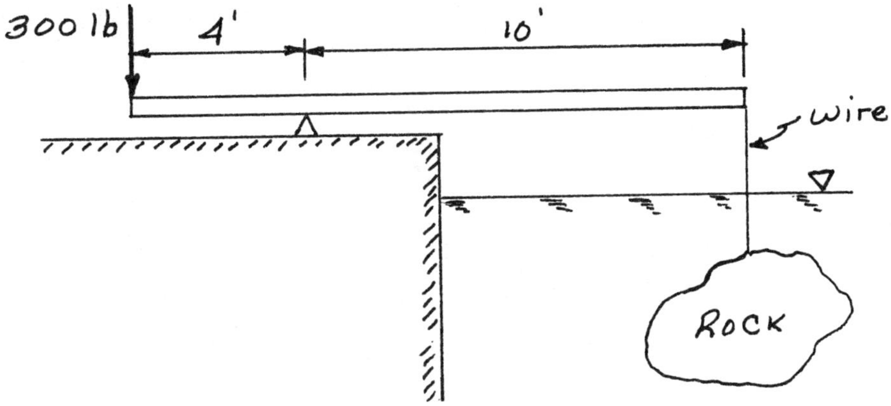

****

FBD of beam:

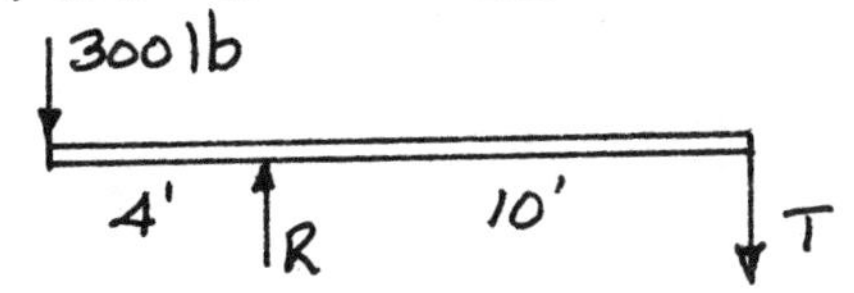

$$\oplus\curvearrowright \quad \Sigma M_R = 0$$

$$10'(T) - 300\,\text{lb}(4') = 0$$

$$T = 120\,\text{lb}$$

FBD of rock:

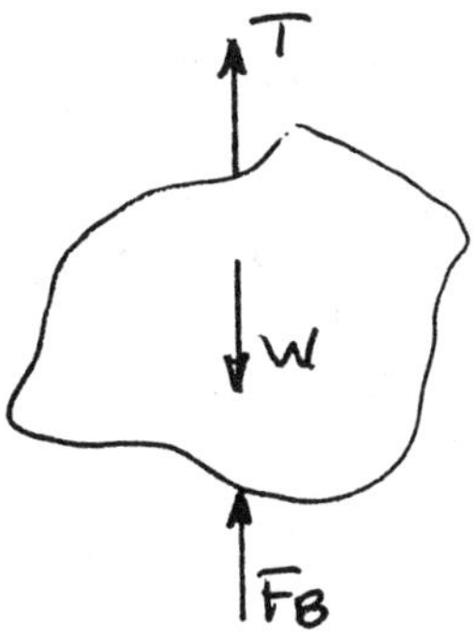

$$\oplus\uparrow \Sigma F_y = 0 \quad \Rightarrow \quad T + F_B - W = 0$$

let V = volume of rock in ft^3

$$120\,\text{lb} + \left(62.4 \tfrac{\text{lb}}{\text{ft}^3}\right)(V) - \left(2.15 \times 62.4 \tfrac{\text{lb}}{\text{ft}^3}\right)(V) = 0$$

$$120 + 62.4\,V - 134.16\,V = 0$$

$$V = 1.67\,\text{ft}^3$$

$$\therefore W = \left(2.15 \times 62.4 \tfrac{\text{lb}}{\text{ft}^3}\right)\left(1.67\,\text{ft}^3\right)$$

$$\underline{\underline{W = 224.3\,\text{lb}}}$$

2-28

A block of wood is floating in water. If γ_{wood} = 8.5 kN/m^3, what is the submerged depth, d?

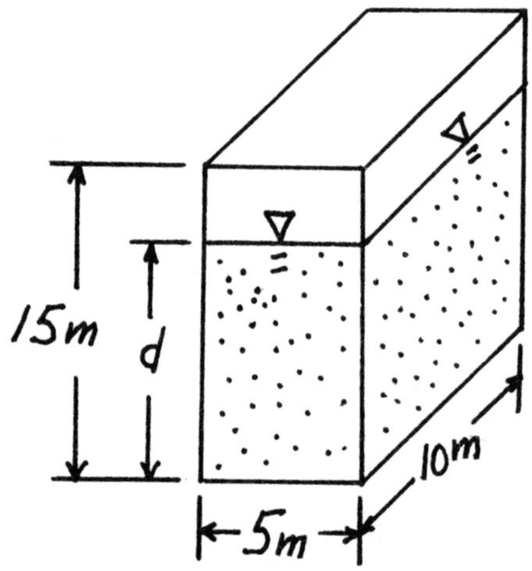

****

$$F_{buoyancy} = Weight_{wood}$$

$$\gamma_{water}(\text{Volume displaced}) = (\gamma_{wood})(\text{Volume}_{wood})$$

$$\left(9.81 \frac{kN}{m^3}\right)(d)(5m)(10m) = \left(8.5 \frac{kN}{m^3}\right)(15m)(10m)(5m)$$

$$d = \frac{\left(8.5 \frac{kN}{m^3}\right)(15m)}{9.81 \frac{kN}{m^3}}$$

$$\underline{\underline{d = 13.0\,m}}$$

2-29

Prove that the block of wood (s.g.= 0.8) in the Figure is not in stable floatation.

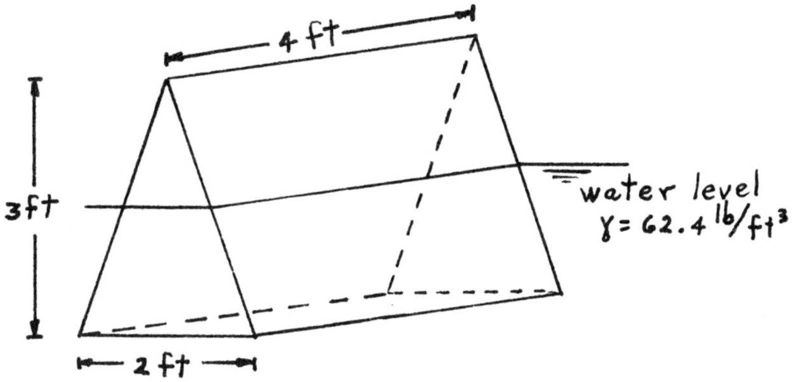

****

block volume $= \frac{1}{2}\, 2ft \times 3ft \times 4ft = 12\ ft^3$

block weight $= 12 ft^3 \times 0.8 \times 62.4 \frac{lb}{ft^3} = 599.04 lb$

volume of water displaced $= \dfrac{599.04 lb}{62.4\ lb/ft^3} = 9.6\ ft^3$

water level over bottom

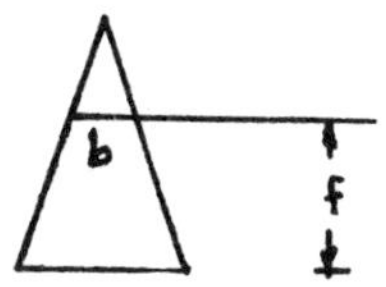

$$\frac{3ft}{2ft} = \frac{3ft - f}{b}$$

$$b = 2 - \frac{2}{3} f$$

$$9.6 ft^3 = \left(\frac{2+b}{2}\right) f \times 4$$

$$19.2 = 8f + 4bf$$
$$19.2 = 8f + 4f\left(2 - \frac{2}{3}f\right)$$
$$19.2 = 8f + 8f - \frac{8}{3}f^2$$
$$57.6 = 48f - 8f^2$$
$$f^2 - 6f + 7.2 = 0$$

$$f = \frac{+6 \pm \sqrt{36 - 28.8}}{4}$$

~~$f_1 = 4.34$~~ no

$f_2 = 1.658\ ft$

$$b = 2 - \frac{2}{3} \times 1.658$$
$$b = 0.895\ ft$$

All distances from bottom:

Center of gravity $= 1$ ft

Center of buoyancy:

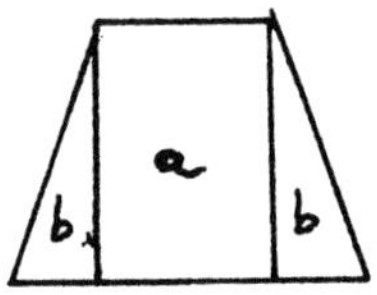

Area	Distance	Moment	
$(1.658 \times 0.894) \times$	0.8292	1.229	(a)
$2 \times (\frac{1}{2} \times 0.5525 \times 1.658) \times$	0.5527	0.506	(b)
2.3997		1.735	

$$\bar{y} = \frac{1.735}{2.3997} = 0.723 \text{ ft}$$

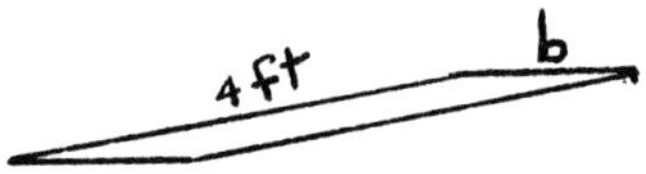

$$BM = \frac{I}{\forall_d}$$

$$I = \frac{BH^3}{12} = \frac{4\text{ft} * (0.895\text{ft})^3}{12} = 0.239 \text{ ft}^4$$

$$BM = \frac{0.239 \text{ ft}^4}{9.6 \text{ ft}^3} = 0.0249 \text{ ft} \quad (\text{distance B to M})$$

$$1\text{ft} - 0.723\text{ft} = 0.277\text{ft} \, (\text{distance B to G})$$

Thus M is closer to B than G

making it <u>unstable</u>

2-30

A stopper is placed over a drain and the tank is filled as shown. Determine the force that must be applied to remove the stopper.

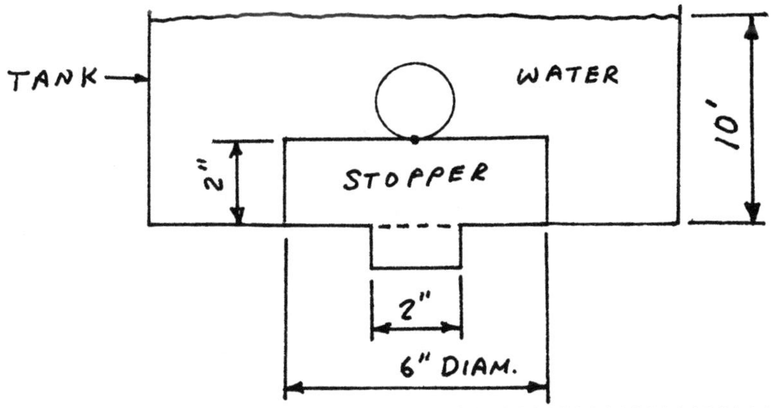

****

IT IS ASSUMED THAT THE OUTER EDGE OF THE STOPPER HAS HYDROSTATIC UPLIFT FORCES. THE TOTAL FORCE IN THIS REGION CAN BE REPRESENTED BY THE BUOYANT FORCE, F_B

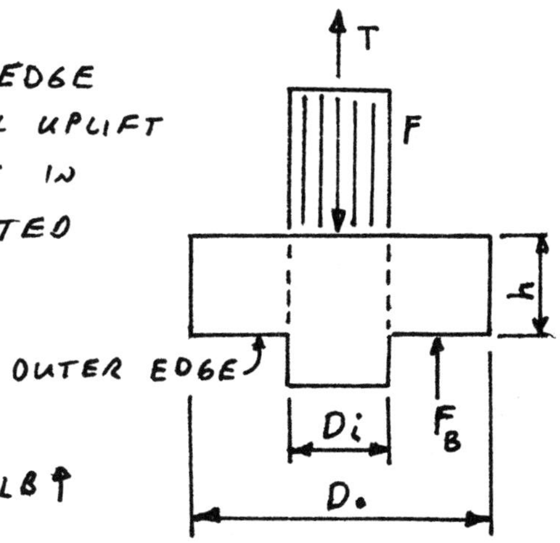

$$F_B = \gamma_{H_2O} \, \forall_{DISP.} = \gamma \frac{\pi}{4}(D_o^2 - D_i^2)\, h$$

$$F_B = \left(62.4 \frac{LB}{FT^3}\right) \frac{\pi}{4}\left(\left(\frac{6}{12}\right)^2 - \left(\frac{2}{12}\right)^2\right)\left(\frac{2}{12}\right) = 1.82 \text{ LB} \uparrow$$

THE INTER SECTION HAS UPLIFT FORCES.

$$F = \gamma h A = \left(62.4 \frac{LB}{FT^3}\right)\left(10 - \frac{2}{12}\right) \frac{\pi}{4}(D_i)^2 = 62.4(9.83) \frac{\pi}{4}\left(\frac{2}{12}\right)^2$$

$$F = 13.38 \text{ LB} \downarrow$$

SOLVE FOR REQUIRED FORCE, T

$$\overset{+}{\uparrow} \Sigma F_y = T + F_B - F = T + 1.82 - 13.38 \text{ LB}$$

$$T = 13.38 - 1.82 = 11.56 \approx \underline{\underline{11.6 \text{ LB} \uparrow}}$$

ACCELERATION ON FLUID MASSES

2-31

Determine the amount of water spilled if the two foot wide container is accelerated at 80 ft/s^2 to the right.

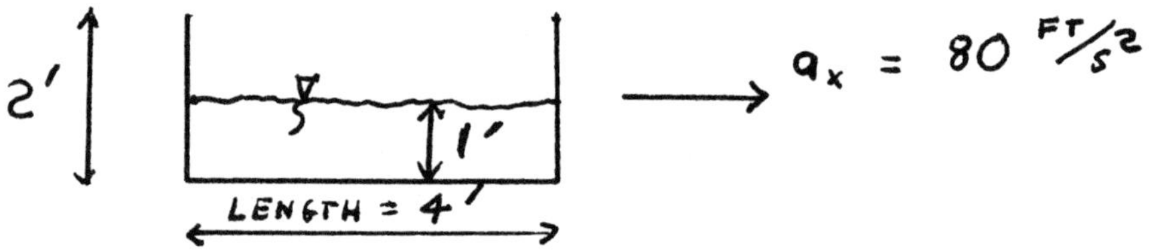

****

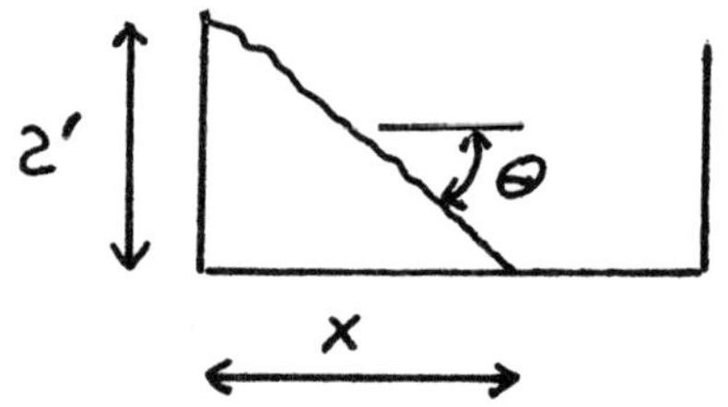

$$\tan\theta = -\frac{a_x}{a_y + g} = \frac{-80}{0 + 32.2} = -2.48$$

SO $-2.48 = -\frac{2}{x} \Rightarrow x = \frac{2}{2.48} = 0.807'$

INITIAL VOLUME $= 1 \times 2 \times 4 = 8$ FT^3

Final VOLUME $= \frac{1}{2}(2 \times .807) \times 2 = 1.6$ Ft^3

VOLUME SPILLED $= 8 - 1.6 = \underline{6.4 \text{ FT}^3}$

2-32

Water is contained in an open vessel up to a depth of one foot as indicated by the dotted horizontal line. Determine the maximum rotational speed (revolutions per minute, rpm) of the vessel around its vertical axis such that no spilling occurs and the gage pressure at the center of the bottom of the vessel is zero. The vessel has a diameter of 2 ft and is 2 ft high.

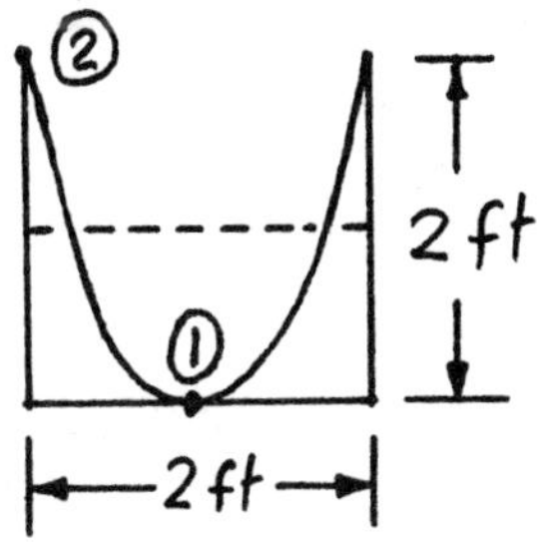

****

This is the case of a forced vortex. The piezometric head differential equation for it, is:

$$\left(\frac{p_2}{\rho g}+z_2\right)-\left(\frac{p_1}{\rho g}+z_1\right)=\frac{\omega^2}{2g}\left(r_2^2-r_1^2\right)$$

Here, $p_1 = p_2 = 0$, $z_1 = 0$, $z_2 = 2$ ft, $r_1 = 0$, $r_2 = 1$ ft, $g = 32.174$ ft/sec^2. After substitution and simplification, one obtains

$$\omega = \text{rotational speed} = \sqrt{\frac{2g\,z_2}{r_2^2}} = \sqrt{\frac{2\times 32.174\times 2}{1}}$$

$\omega = 11.344$ rad/sec, speed $= \dfrac{\omega \times 60}{2\pi}$ rpm

or, speed $= \dfrac{11.344\times 60}{2\pi} = 108.3$ rpm

2-33

The tanker truck shown below is forced to make an emergency stop on the highway, decelerating at a uniform rate of 1.5 m/s^2. What is the maximum pressure (in kPa) exerted on the tank compartment if it is carrying oil with a specific gravity of 0.86?

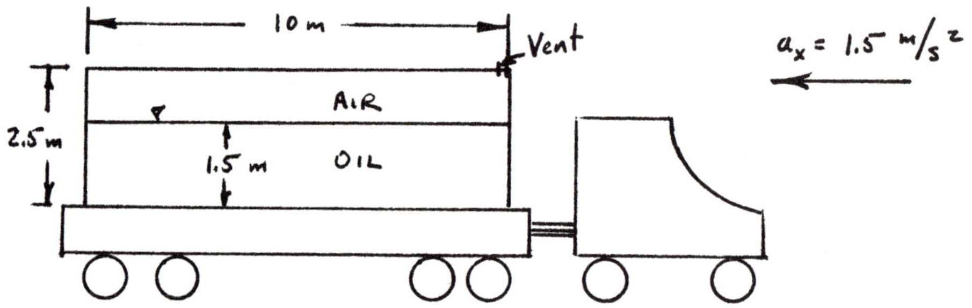

****

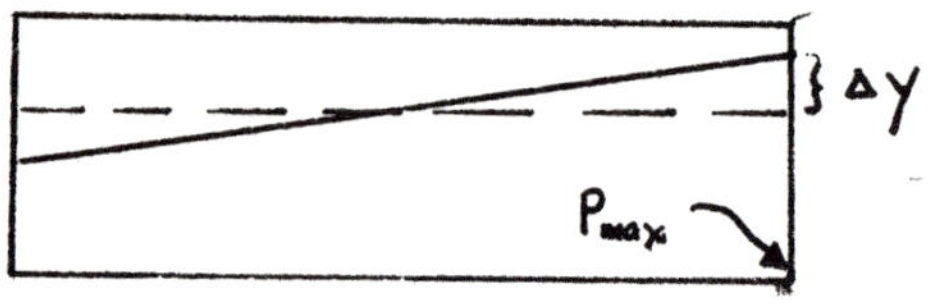

$$dp = \frac{\partial p}{\partial x}dx + \frac{\partial p}{\partial y}dy$$

$$= -\rho a_x\,dx - (\rho a_y^{\,0} + \rho g)\,dy$$

$$dp = 1.5\rho\,dx - 9.8\rho\,dy$$

Free surface deflected slope: $dp = 0$

$$9.8\rho\,dy = 1.5\rho\,dx$$

$$\frac{dy}{dx} = \frac{1.5}{9.8} = 0.153$$

$$\Delta y = 0.153(5\,\text{m}) = 0.765\,\text{m}$$

$$P_{max} = (0.765 + 1.5)_{m}(0.86)(9.8\ \text{kN/m}^3)$$

$$\underline{\underline{P_{max} = 19.1\ \text{kPa}}}$$

2-34

A rectangular tank is filled with water to a depth of four feet. The tank is accelerated to the right. Determine the magnitude of the pressure in all corners of the tank. Calculate the force on the ends of the tank.

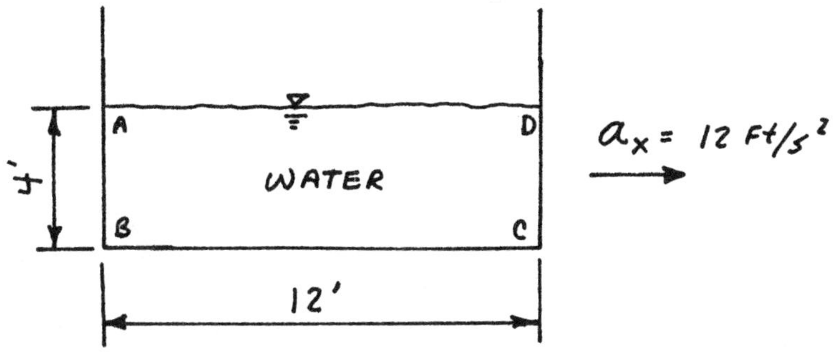

****

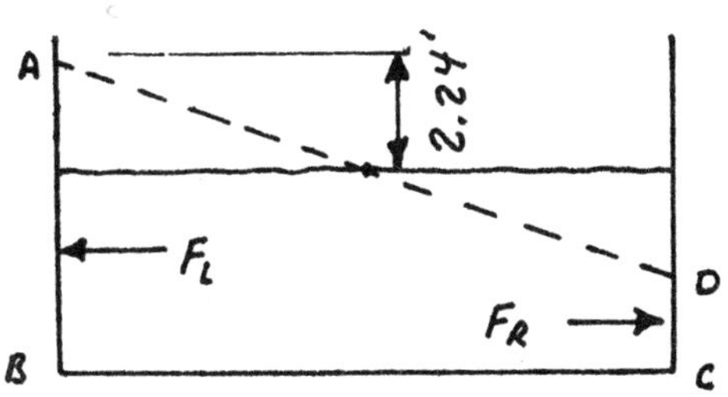

CALCULATE SLOPE OF SURFACE

$$\frac{dz}{dx} = -\left(\frac{a_x}{g + a_z}\right) = -\left(\frac{12}{32.2+0}\right)$$

$$\frac{dz}{dx} = -0.373$$

WATER PIVOTS ABOUT CENTER AND ELEVATION CHANGES IN A 6 FOOT DISTANCE FROM CENTER $\quad 6 \times 0.373 = 2.24'$

WATER DEPTH AT B = $4 + 2.24 = 6.24'$
WATER DEPTH AT C = $4 - 2.24 = 1.76'$

$P_A = 0$ psf $\qquad P_D = 0$ psf.

$$P_B = \gamma_{H_2O}\, h = 62.4\ \frac{LB}{ft^3}\,(6.24') = \underline{\underline{389.4\ psf}}$$

$$P_C = \gamma_{H_2O}\, h = 62.4\ \frac{LB}{Ft^3}\,(1.76\ ft) = \underline{\underline{109.8\ psf}}$$

CALCULATE FORCES ON ENDS.

$$F_L = \gamma h_c A = (62.4)\left(\frac{6.24}{2}\right)(6.24 \times 1) = \underline{\underline{1215}}\ LB$$

$$F_R = \gamma h_c A = (62.4)\left(\frac{1.76}{2}\right)(1.76 \times 1) = \underline{\underline{96.6}}\ LB$$

2-35

A tank containing a fluid of specific gravity of 0.8 is being pulled at a constant acceleration of 10 ft/sec². Calculate the total force on the wall AB of the tank.

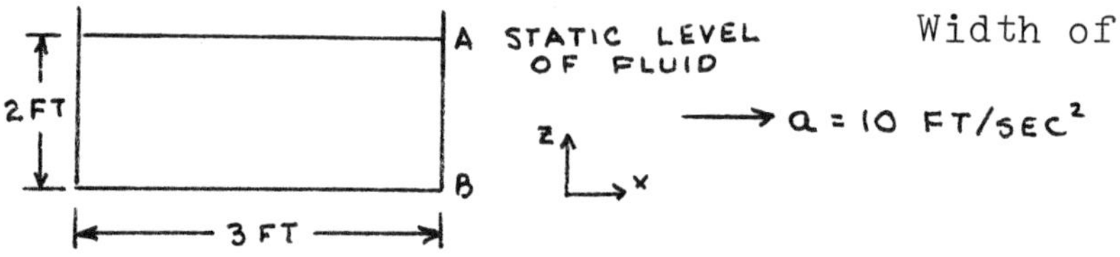

Width of tank= 3 ft

****

THE EQUATION GOVERNING THE PRESSURE IN THE FLUID IS

$$\vec{\nabla} p = -\gamma \vec{K} - \vec{a}$$

WHERE $\vec{a}$ IS THE ACCELERATION VECTOR

$$\frac{\partial p}{\partial x}\vec{i} + \frac{\partial p}{\partial z}\vec{K} = -a\vec{i} - \gamma \vec{K}$$

$$\frac{\partial p}{\partial x} = -a\frac{\gamma}{g} \longrightarrow p = -ax\frac{\gamma}{g} + f(z)$$

$$\frac{\partial p}{\partial z} = -\gamma \longrightarrow p = -\gamma z + g(x)$$

$$-ax\frac{\gamma}{g} + f(z) = -\gamma z + g(x) \longrightarrow f(z) = -\gamma z + C \,;\; g(x) = -ax\frac{\gamma}{g} + C$$

$$p(x,z) = -\gamma z - ax\frac{\gamma}{g} + C$$

IF THE ORIGIN IS CHOSEN AT THE CENTER OF THE BOTTOM OF THE TANK. THEN AT $x = 0$, $z = h = 2$ FT, $p = 0$. HENCE

$$0 = -\gamma h + C \qquad \text{OR} \qquad C = \gamma h$$

$$P(x,z) = \gamma(h-z) - ax\frac{\gamma}{g}$$

THE FREE SURFACE DEFINED BY $p = 0$. THUS

$0 = \gamma(h-z) - ax\frac{\gamma}{g}$ DEFINES THE FREE SURFACE

$\gamma = 0.8(62.4 \text{ LB/FT}^3) \qquad h = 2$

THE WALL AB IS DEFINED BY $x = 1.5$ FT. THUS FOR $x = 1.5$ FT THE FREE SURFACE BEGINS AT A VALUE OF z GIVEN BY

$$0 = 0.8(62.4 \text{ LB/FT}^3)(2 \text{ FT} - z) - 10 \text{ FT/SEC}^2 (1.5)\frac{(0.8)(62.4)}{(32.2)}$$

$$z = 1.53 \text{ FT}$$

THE PRESSURE FOR $x = 1.5$ FT IS

$$P(1.5, z) = 49.92(2-z) - (1.5)(10)\frac{49.92}{32.2} = 76.59 - 49.92\,z$$

THE FORCE IS GIVEN BY

$$F_{AB} = \int p\,dA = \int_0^{1.53} (76.59 - 49.92\,z)(3\text{ FT})\,dz$$

$$= 3\left(76.59\,z - 49.92\,\frac{z^2}{2}\right)\Big|_{z=0}^{z=1.53\text{ FT}}$$

$$= 176.3 \text{ LBS}$$

2-36

A tanker is filled with water up to three quarters of its height. The length and height of this tanker are 26 and 8 ft, respectively. If this tanker is accelerated at a constant rate of 10 ft/s^2, determine whether the water will rise to the top of this tanker.

****

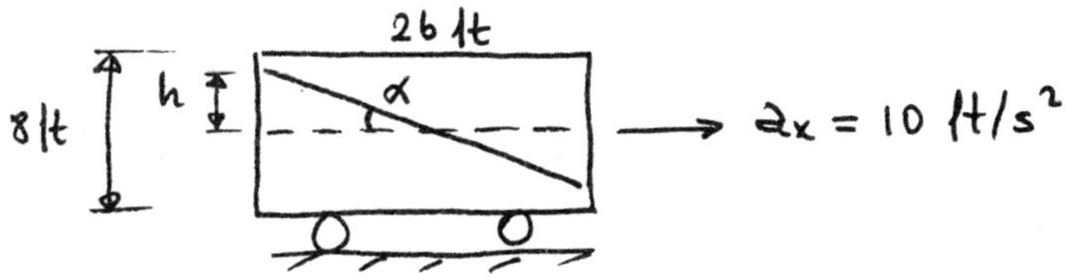

$$\tan\alpha = \frac{a_x}{g} = \frac{10}{32.2}$$

also,

$$\tan\alpha = \frac{h}{26/2} \longrightarrow h = \frac{26}{2} \times \tan\alpha = 13 \times \frac{10}{32.2} = 4.04 \text{ ft}.$$

$$h = 4.04 \text{ ft} > \frac{8}{2} = 4 \text{ ft}$$

Therefore, water will touch the the top surface of the tanker.

2-37

A closed cylindrical tank is rotated about its axis at 100 rev/min. The tank is filled with alcohol to a depth of three meters. A pressure of 20 kPa is applied to the air within the tank. Calculate the pressure on the bottom of the tank at the center and outer edge. What is the total uplift force on the lid of the tank? Tank is 4 m high and 2 m in diameter.

****

DEFINE THE WATER SURFACE.

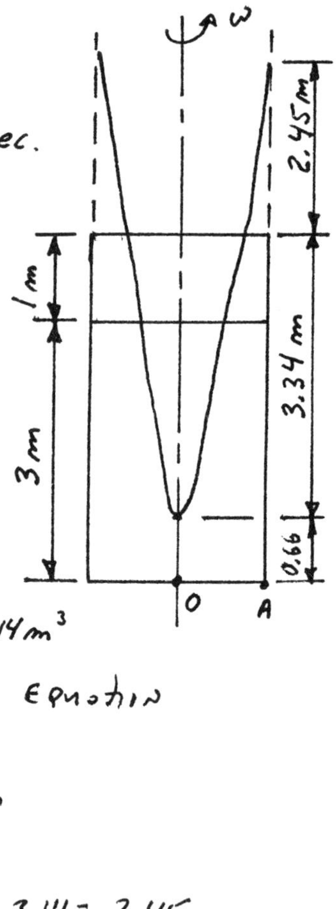

$$\omega = \frac{100 \text{ REV}}{\text{MIN}} \; \frac{2\pi \text{ RAD}}{\text{REV}} \; \frac{\text{MIN}}{60 \text{ SEC}} = 10.47 \text{ RAD/sec.}$$

$$z = \frac{\omega^2 r^2}{2g} = \frac{(10.47 \text{ RAD/sec})^2 (1 m)^2}{(2)(9.81 \text{ m/s}^2)} = 5.59 \text{ m}$$

SINCE $z/2$ IS GREATER THAN THE 1 m FREE AIR SPACE IT IS NECESSARY TO KEEP AIR SPACE CONSTANT.

$$V_{AIR} = \frac{\pi D^2}{4} h = \frac{\pi (2)^2}{4} (1) = 3.14 \text{ m}^3$$

$$V_{parabola} = \frac{1}{2}(\pi r^2 h) = \frac{1}{2} \pi r^2 \left(\frac{\omega^2 r^2}{2g}\right) = 3.14 \text{ m}^3$$

radius, $r = 0.773$ m PLUG INTO "z" EQUATION

$$z = \frac{\omega^2 r^2}{2g} = \frac{(10.47)^2 (0.773)^2}{(2)(9.81)} = 3.34 \text{ m}$$

REMAINING HEIGHT OF PARABOLA = 5.59 − 3.14 = 2.45 m

$$P_0 = \gamma_{AL} h + P_{AIR} = S.G.\, \gamma_{H_2O @ 20°C}\, h + P_{AIR}$$

$$= (0.79)(9.789 \text{ kN/m}^3)(0.66 \text{ m}) + 20 \text{ kPa} = 5.1 + 20 = \underline{\underline{25.1 \text{ kPa}}}$$

$$P_A = \gamma_{AL} h + P_{AIR} = S.G.\, \gamma_{H_2O}\, h + P_{AIR}$$

$$= (0.79)(9.789 \text{ kN/m}^3)(6.45 \text{ m}) + 20 \text{ kPa} = 49.9 + 20 = \underline{\underline{69.9 \text{ kPa}}}$$

THE UPLIFT FORCE ON THE LID IS EQUAL TO BUOYANT FORCE OF PROJECTED VOLUME ABOVE LID PLUS AIR PRESSURE TIME AREA OF LID.

$$\forall_{AL} = \forall_c - \forall_p + \forall_{AIR} = \pi r^2 h - \tfrac{1}{2}\pi r^2 h + \tfrac{1}{2}\pi r^2 h$$

$$= \pi (1)^2 (2.45) - \tfrac{1}{2}\pi (1)^2 (5.59) + \tfrac{1}{2}\pi (0.773)^2 (3.34)$$

$$= 7.70 - 8.78 + 3.13 = 2.05 \ m^3$$

$$F_{UPLIFT} = F_B + F_p = \gamma_{AL} \forall_{AL} + PA$$

$$= (0.79)(9.789 \ kN/m^3)(2.05 m^3) + (20 \ KPa)(\pi)(1)^2$$

$$= 15.85 + 62.83 = \underline{\underline{78.68 \ kN}}$$

2-38

The figure shows a two dimensional stream tube, drawn to scale. The flow rate is 40 cfs/ft perpendicular to the plane of the sketch. Determine approximate values of the normal and tangential accelerations of a fluid particle at A. State any approximations that you make. Scale the figure as necessary.

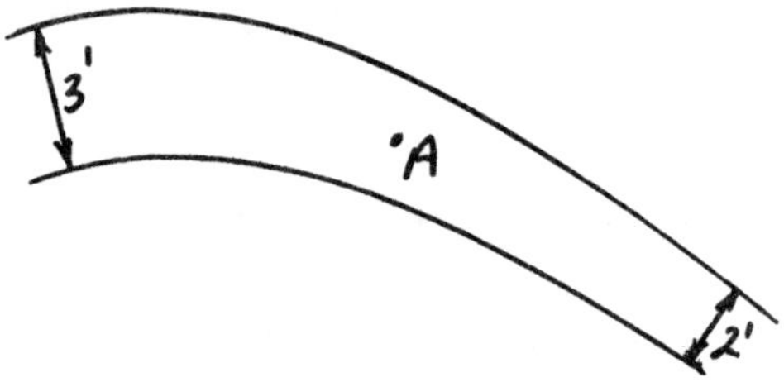

****

$$V_1 \approx \frac{40}{2.75} = 14.55 \text{ ft/sec}$$

$$V_2 \approx \frac{40}{2.3} = 17.39 \text{ ft/sec}$$

$$V_A \approx \frac{40}{2.6} = 15.38 \text{ ft/sec}$$

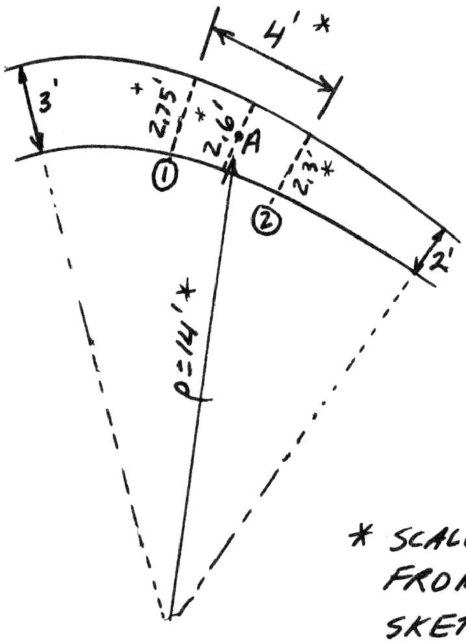

* SCALED FROM SKETCH

$$a = \frac{dV}{dt}, \quad V = \frac{ds}{dt}$$

ELIM dt: $a_t = V\dfrac{dV}{ds}$

$$a_t = V\frac{dV}{ds} \approx V\frac{\Delta V}{\Delta S} = 15.38\frac{(17.39 - 14.55)}{4} = \boxed{10.9 \text{ ft/s}^2}$$

ANS.

$$a_N = \frac{V^2}{\rho} = \frac{(15.38)^2}{14} = \boxed{16.9 \text{ ft/s}^2} \quad \text{ANS}$$

Note: Answers may vary depending on scaling.

3 FLUIDS IN MOTION

VELOCITY DISTRIBUTION AND RATE OF FLOW

3-1

Velocity distribution in a pipe is given by $U = 2(1-(r/R)^2)$ ft/s. Pipe diameter is given as 2 ft. Determine the volumetric flow rate and the mean velocity in this pipe.

****

Radius of the pipe is $R = 1$ ft.

Volumetric flow rate:

$$Q = 2\pi \int_0^1 u \cdot r \cdot dr = 2\pi \int_0^1 2\left(1-\left(\frac{r}{R}\right)^2\right) dr$$

$$= 4\pi \int_0^1 (1-r^2) r\,dr = 4\pi \int_0^1 (r-r^3)\,dr$$

Integration gives

$$Q = 4\pi \left(\frac{r^2}{2} - \frac{r^4}{4}\right)\Big|_0^1 = 4\pi\left(\frac{1}{2}-\frac{1}{4}\right) = \pi \text{ ft}^3/\text{s}$$

Mean flow velocity:

$$\bar{U} = \frac{Q}{A} = \frac{Q}{\pi\left(\frac{d^2}{4}\right)} = \frac{\pi}{\pi\left(\frac{2^2}{4}\right)} = 1 \text{ ft/s}.$$

3-2

A fluid in a wide channel has a velocity profile as shown.
Calculate q (the flow rate in cfs/ft of width) and the mean velocity.

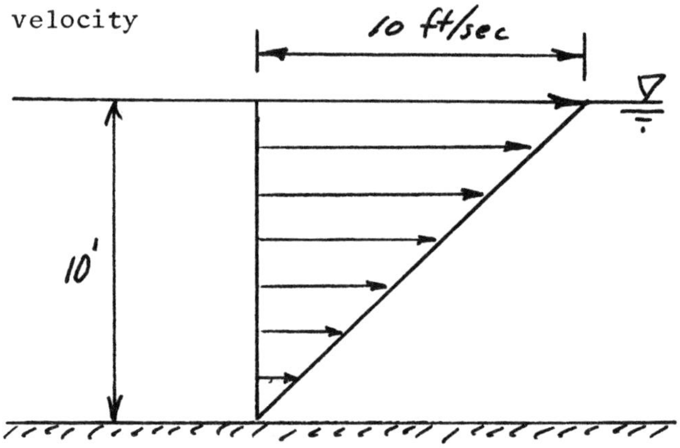

****

Equation for velocity profile is $v = y$

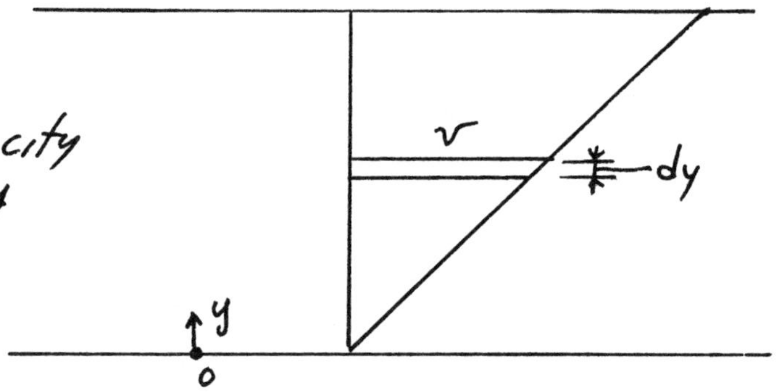

$$q = \int_0^{10} v\,dy = \int_0^{10} y\,dy = \frac{y^2}{2}\Big]_0^{10} = \boxed{50 \text{ cfs/ft width}}$$

$$\text{Avg } v = \frac{q}{A} = \frac{50}{10(1)} = \boxed{5 \text{ ft/sec}}$$

-OR- (in this simple case):

$$q = \frac{1}{2} bh = \frac{1}{2}(10)(10) = 50 \text{ cfs/ft width}$$

and $$\text{Avg } v = \frac{10}{2} = 5 \text{ ft/sec}$$

3-3

The volume flow rate through the 8 cm diameter pipe is 600 cm^3/s. The flow is uniform at station 1 and parabolic at station 2. Find v_1 and C.

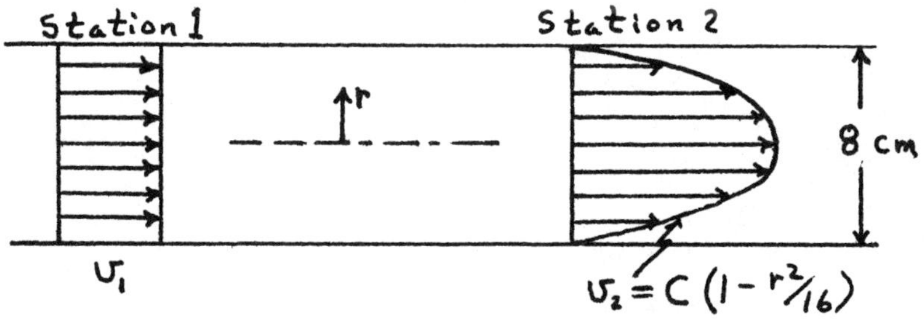

****

from continuity,
$$v_1 = \frac{Q}{A_1} = \frac{600 \ cm^3/s}{\pi(4)^2 \ cm^2} = 11.94 \ cm/s$$

for station 2

$$dA = 2\pi r\,dr$$

$$Q = \int v_2\,dA = \int_0^4 C(1 - r^2/16)\,2\pi r\,dr = 2\pi C\int_0^4 \left(r - \frac{r^3}{16}\right)dr$$

$$Q = 2\pi C\left[\frac{r^2}{2} - \frac{r^4}{64}\right]_0^4 = 2\pi C\left(\frac{4^2}{2} - \frac{4^4}{64}\right) = 8\pi C$$

$$Q = 600 = 8\pi C \qquad C = \frac{600}{8\pi} = 23.88 \ cm/s$$

Note that C is the velocity at the center, and it is exactly twice that of the average ($v_1 = 11.94$ cm/s).

3-4

A velocity field is given by

$$\vec{v} = (t+1)\ (x^2 z\ \vec{i}\ -2xyz\ \vec{j})$$

where x,y,z are Cartesian coordinates and $\vec{i}$, $\vec{j}$, and $\vec{k}$ are unit vectors in these coordinate directions.

a. Is the flow field steady ? Uniform ?
b. Is the flow field incompressible ? Irrotational ?
c. At t=0 what is the acceleration of a fluid particle located at (1,3,2)?
d. Find the volume flux at t=0 through the triangle bounded by the verticies (1,0,0), (1,1,1), and (1,0,2).

****

a) THE FLOW FIELD IS UNSTEADY BECAUSE THE FLOW PROPERTIES DEPEND UPON TIME. THE FLOW FIELD IS NONUNIFORM BECAUSE THE FLOW PROPERTIES VARY WITH SPATIAL POSITION.

b) THE FLOW IS INCOMPRESSIBLE IF $\text{div}\ \vec{V} = 0$

$$\text{div}\ \vec{v} = \frac{\partial V_x}{\partial x} + \frac{\partial V_y}{\partial y} + \frac{\partial V_z}{\partial z} = \frac{\partial}{\partial x}\left[(t+1)(x^2 z)\right] + \frac{\partial}{\partial y}\left[(t+1)(-2xyz)\right] + \frac{\partial}{\partial z}(0)$$

$$= (t+1)\left[2xz - 2xz\right] = 0 \longrightarrow \text{THE FLOW IS INCOMPRESSIBLE}$$

THE FLOW IS IRROTATIONAL IF $\text{curl}\ \vec{V} = 0$

$$\text{curl}\ \vec{v} = \begin{vmatrix} \vec{i} & \vec{j} & \vec{k} \\ \frac{\partial}{\partial x} & \frac{\partial}{\partial y} & \frac{\partial}{\partial z} \\ V_x & V_y & V_z \end{vmatrix}$$

$$= \vec{i}\left(\frac{\partial V_y}{\partial x} - \frac{\partial V_x}{\partial y}\right) - \vec{j}\left(\frac{\partial V_z}{\partial x} - \frac{\partial V_x}{\partial z}\right) + \vec{k}\left(\frac{\partial V_z}{\partial y} - \frac{\partial V_y}{\partial z}\right)$$

$$= (t+1)\left[-2xy\ \vec{i} + 2x\ \vec{j} - 2yz\ \vec{k}\right]$$

$$\neq 0 \longrightarrow \textit{THE FLOW IS NOT IRROTATIONAL}$$

c) $$\vec{a} = \frac{D\vec{V}}{Dt} = \frac{\partial \vec{V}}{\partial t} + V_x \frac{\partial \vec{V}}{\partial x} + V_y \frac{\partial \vec{V}}{\partial y} + V_z \frac{\partial \vec{V}}{\partial z}$$

$$= (x^2 z\ \vec{i} - 2xyz\ \vec{j}) + (t+1)(x^2 z)\left[(t+1)2xz\ \vec{i} - (t+1)2yz\ \vec{j}\right]$$
$$- (t+1)^2 (2xyz)(-2xz\ \vec{j})$$

AT (1,3,2) WHEN $t=0$ $\quad \vec{a} = 10\vec{i} + 12\vec{j}$

d) $Q = \int \vec{V}\cdot\vec{n}\ dA$ WHERE $\vec{n}$ IS THE NORMAL VECTOR

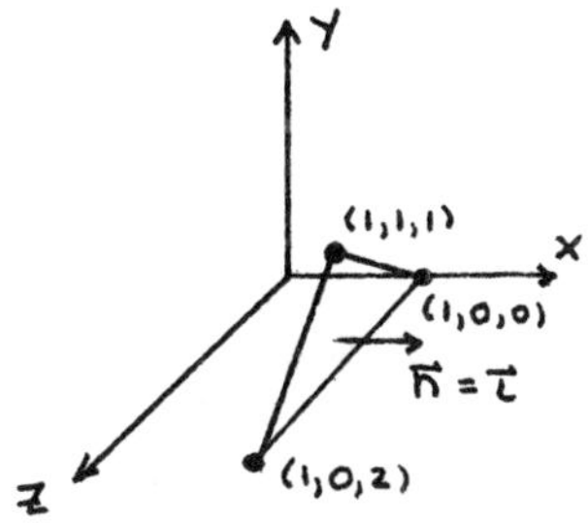

$\vec{v} \cdot \vec{n} = (t+1)x^2 z$

AT $t=0$ $\quad \vec{v} \cdot \vec{n} = x^2 z$

HOWEVER OVER THE SURFACE WHERE THE INTEGRAL IS BEING EVALUATED $x=1$. THUS

$$Q = \int z \, dA \quad \text{OVER THE TRIANGLE}$$

HOWEVER $\int z \, dA$ IS THE z-COORDINATE OF THE CENTROID OF THE TRIANGLE TIMES THE AREA OF THE TRIANGLE

$$Q = (1)(1) = 1$$

3-5

Water flows between two parallel plates as shown.
If the velocity profile is $v_z = v_{max}\,[1 - (y/B)^2]$, find the volumetric flowrate Q for width W (in the x-direction).
What is the average velocity $\langle v_z \rangle$?

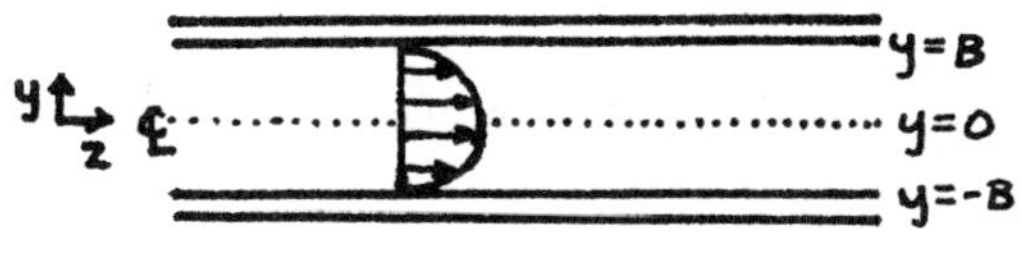

****

$$Q = \int v_z \, dA = \int_0^W \int_{-B}^{B} v_{max}\left[1-\left(\tfrac{y}{B}\right)^2\right] dx\,dy = v_{max} W \int_{-B}^{B} \left[1-\left(\tfrac{y}{B}\right)^2\right] dy$$

$$Q = v_{max} W \left[y - \frac{y^3}{3B^2}\right]_{-B}^{B} = v_{max} W \left[B - \frac{B}{3} - (-B) + \frac{(-B)^3}{3B^2}\right] = \frac{4}{3} B W v_{max}$$

$$\langle v_z \rangle = \frac{Q}{A} = \frac{\frac{4}{3} B W v_{max}}{2BW} = \frac{2 v_{max}}{3}$$

CONTINUITY EQUATION

3-6

If twice as much water discharges through the bottom outlet as the top for the reservoir shown, determine the density (in kg/m^3) of the final discharge flow.

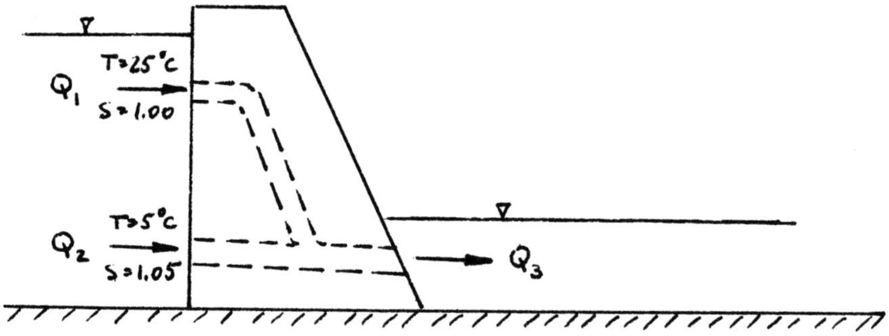

****

Apply continuity for mass flow rates:

$$\rho_1 Q_1 + \rho_2 Q_2 = \rho_3 Q_3$$

$Q_1 = Q$ $\qquad \rho_1 = 997 \text{ kg/m}^3$

$Q_2 = 2Q$ $\qquad \rho_2 = 1.05(1000) = 1050 \text{ kg/m}^3$

$Q_3 = 3Q$

substitute:

$$997Q + 1050(2Q) = \rho_3 (3Q)$$

$$\therefore \rho_3 = \frac{997 + 2100}{3} = \underline{1032 \text{ kg/m}^3}$$

3-7

The water tank shown below is being filled through sections 1 and 2 while water is leaving at section 3. The velocity of the fluid entering at 1 is 10 ft/sec. The exit velocity of the fluid is approximated by Torricelli's Law. What should the flow rate through 2 be so that the height of liquid in the tank increases at a constant rate of 1 ft/sec? The initial height of liquid is 3 ft.

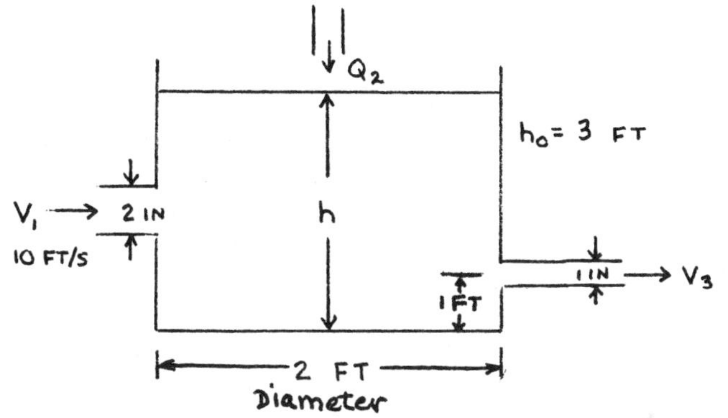

Notes:
(1) Q_2 may be a function of time.
(2) Torricelli's Law $V = \sqrt{2g(h-1)}$

****

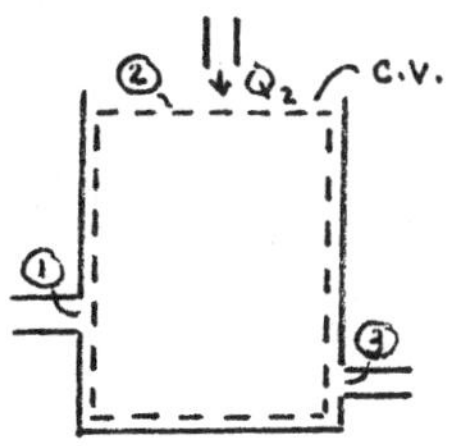

ASSUMPTIONS:
INCOMPRESSIBLE
ONE-DIMENSIONAL APPROXIMATION AT INLETS AND OUTLETS

CONTINUITY

$$\frac{\partial}{\partial t}\int_{C.V.} \rho \, d\forall + \int_{C.S.} \rho \vec{v}\cdot\vec{n}\, d\sigma = 0$$

OR

$$\frac{d\forall}{dt} + \int_{①} \vec{v}_1\cdot\vec{n}_1\, d\sigma_1 + \int_{②} \vec{v}_2\cdot\vec{n}_2\, d\sigma_2 + \int_{③} \vec{v}_3\cdot\vec{n}_3\, d\sigma_3 = 0$$

BUT

$$\forall = A_{TANK}\, h \longrightarrow \frac{dv}{dt} = A_{TANK}\frac{dh}{dt}$$

THUS

$$A_{TANK}\frac{dh}{dt} - V_1 A_1 - Q_2 + V_3 A_3 = 0$$

$$Q_2 = A_{TANK}\frac{dh}{dt} - V_1 A_1 + V_3 A_3$$

OR $$Q_2 = \pi (1\,FT)^2 (1\,FT/SEC) - 10\,FT/SEC\,(1/12\,FT)^2 + \left(\frac{1}{24}\,FT\right)^2 \pi \sqrt{2g(h-1)}$$

BUT $$\frac{dh}{dt} = 1 \longrightarrow h = t + h_o = 3 + t$$

THEREFORE $$Q_2 = (3.07 + 0.0438\sqrt{2+t})\ FT^3/SEC$$

3-8

Water is flowing in the pipeline system shown below. Fill in the missing spaces in the accompanying table and determine the direction of flow in pipe D.

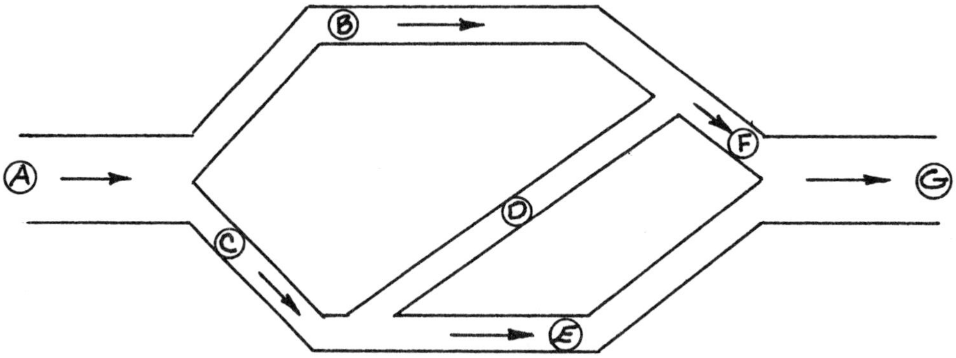

PIPE	FLOW (ft^3/sec)	Velocity (ft/sec)	Area (ft^2)
A	100		20
B		15	
C	60		6
D			10
E		20	
F	20		15
G		4	

****

$$Q_A = V_A A_A \quad \Rightarrow \quad V_A = \frac{100\ ft^3/sec}{20\ ft^2} = 5\ ft/sec$$

$$Q_A = Q_G = 100\ ft^3/sec$$

$$Q_G = V_G A_G \quad \Rightarrow \quad A_G = \frac{100\ ft^3/sec}{4\ ft/sec} = 25\ ft^2$$

$$Q_A = Q_B + Q_C \quad \Rightarrow \quad Q_B = (100 - 60)\ ft^3/sec = 40\ ft^3/sec$$

$$Q_B = V_B A_B \quad \Rightarrow \quad A_B = \frac{40\ ft^3/sec}{15\ ft/sec} = 2.67\ ft^2$$

$$Q_C = V_C A_C \quad \Rightarrow \quad V_C = \frac{60\ ft^3/sec}{6\ ft^2} = 10\ ft/sec$$

$$Q_F = V_F A_F \Rightarrow V_F = \frac{20\ ft^3/sec}{15\ ft^2} = 1.33\ ft/sec$$

$$Q_E + Q_F = Q_G \Rightarrow Q_E = (100-20)\ ft^3/sec = 80\ ft^3/sec$$

$$Q_E = V_E A_E \Rightarrow A_E = \frac{80\ ft^3/sec}{20\ ft/sec} = 4\ ft^2$$

Assume flow in pipe D is to the right:

then: $Q_B + Q_D = Q_F \Rightarrow Q_D = 20 - 40 = -20\ ft^3/sec$

Since Q_D is negative, flow in pipe D must be to the left.

check: $Q_C + Q_D = Q_E \Rightarrow Q_D = 80 - 60 = 20\ ft^3/sec$ O.K.

$$Q_D = V_D A_D \Rightarrow V_D = \frac{20\ ft^3/sec}{10\ ft^2} = 2\ ft/sec$$

Solution:

PIPE	FLOW	VELOCITY	AREA
A	100	5	20
B	40	15	2.67
C	60	10	6
D	20	2	10
E	80	20	4
F	20	1.33	15
G	100	4	25

$Q_D = 20\ cfs$ ↙

3-9

The flow in a 200 mm diameter pipe is distributed to a 50 mm and 150 mm diameter pipes. The velocity in the 200 mm diameter pipe is 2 m/s and the velocity of flow in the 50 mm diameter pipe is 8 m/s. Assume ideal flow and full flow in each pipe. Calculate the total flow in each pipe and the velocity of flow in the 150 mm diameter pipe.

****

$$Q_{200} = AV = \frac{\pi(.2)^2}{4}(2) = \underline{\underline{0.0628\ m^3/s}}$$

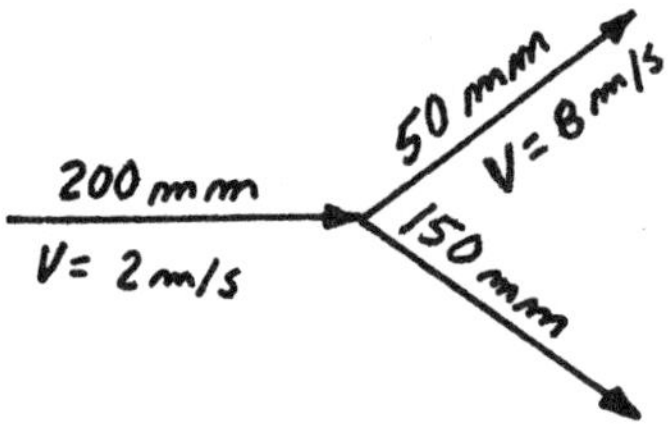

$$Q_{50} = AU = \frac{\pi(.05)^2}{4}(8) = \underline{\underline{0.0157\ m^3/s}}$$

$$Q_{150} = Q_{200} - Q_{50} = 0.0628 - 0.0157 = \underline{\underline{0.0471\ m^3/s}}$$

$$V_{150} = \frac{Q_{150}}{A} = \frac{0.0471}{\frac{\pi(.15)^2}{4}} = \underline{\underline{2.66\ m/s}}$$

3-10

A tank is full of water as shown in the figure below.

(a) How long would it take for a quantity of water equal to the capacity of the tank to flow out, assuming a static head and constant flow rate?

(b) Now assuming no inflow into the tank, how long will it take for the tank to empty?

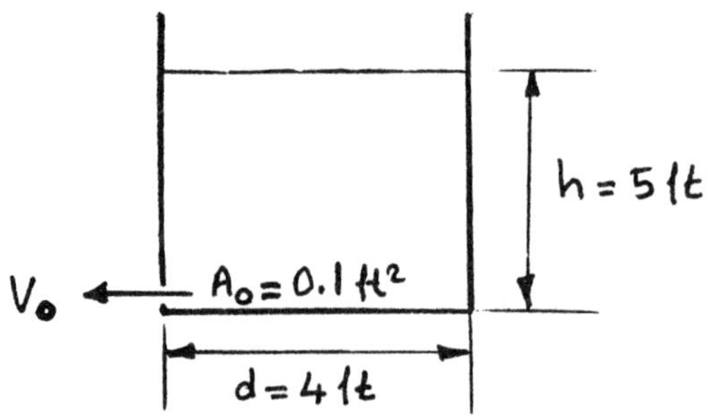

****

Volume of the water in the tank :

$$V_t = A \cdot h = \left(\pi \frac{d^2}{4}\right) \cdot h = \left(\pi \cdot \frac{4^2}{4}\right) 5 = 62.83\ ft^3$$

(a) Water will be considered as incompressible : ρ= const.

Using Bernoulli Eqn., we can obtain the velocity at the orifice as

$$V_o = \sqrt{2gh} = \sqrt{2 \times 32.2 \times 5} = 17.944 \text{ ft/sec}$$

Therefore, volumetric flow rate can be calculated as

$$Q = A_o \cdot V_o = 0.1 \times 17.944 = 1.794 \text{ ft}^3/\text{sec}.$$

$$\text{time required} : \frac{V_t}{Q} = \frac{62.83}{1.794} = 35.02 \text{ seconds}$$

(b) The velocity of the jet changes as the level of water drops. This can be included in the analysis by considering the continuity equation:

$$\sum_{C.\text{Surface}} \rho \vec{V} \cdot \vec{A} = -\frac{d}{dt} \int_{C.\text{Volume}} \rho \, dV$$

$\vec{V}$: velocity vector
dV : control volume element

Applying to this tank, we can get

$$\rho_o V_o A_o = -\frac{d}{dt}(\rho V_t) = -\frac{d}{dt}(\rho A h)$$

and since water is incompressible, ρ = const and additionally $A = \pi d^2/4$ = const. and $V_o = \sqrt{2gh}$ at any time during discharge, we can write

$$A_o \sqrt{2gh} = -A \frac{dh}{dt}$$

This gives, after integration

$$t = \frac{2A}{\sqrt{2g} \cdot A_o} \left(h_{\text{init.}}^{1/2} - h_{\text{last}}^{1/2} \right)$$

$h_{\text{init.}} = h = 5$ ft, $h_{\text{last}} \cong 0$, therefore

$$t = \frac{2 \times \pi \times \frac{4^2}{4}}{\sqrt{2 \times 32.2} \times 0.1} (5)^{1/2} = 70.03 \text{ sec.}$$

Obviously, as the height h decreases, V_o decreases; therefore, time required for discharge increases: $70.03 > 35.02$ seconds.

3-11

Oxygen at 20 psia and 80°F and hydrogen at 18 psia and 85°F are mixed in a chamber. Entry velocities are 10 ft/s and 12 ft/s for the oxygen and hydrogen, respectively. If the mixture leaves the chamber from a single outlet with a velocity of 10 ft/s, determine the density of the mixture at the outlet.

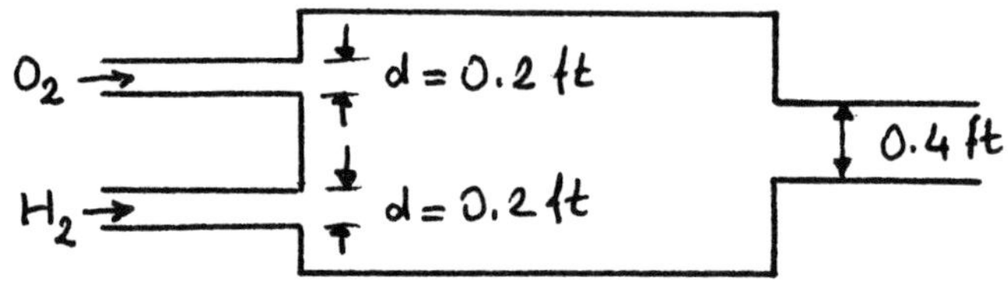

****

For oxygen, assuming perfect gas :

$$\forall = \frac{RT}{P} = \frac{1545.3/32 \times 540}{20 \times 144} = 9.0544 \ \text{ft}^3/\text{lb}_m$$

$$\rho = \frac{1}{\forall} = 0.11 \ \frac{\text{lbm}}{\text{ft}^3} = 0.0034 \ \frac{\text{slug}}{\text{ft}^3}$$

For hydrogen,

$$\upsilon = \frac{RT}{P} = \frac{1545.3/2.016 \times 545}{18 \times 144} = 161.17 \ \text{ft}^3/\text{lb}_m$$

$$\rho = \frac{1}{\upsilon} = 0.0062 \ \frac{\text{lbm}}{\text{ft}^3} = 0.000193 \ \frac{\text{slug}}{\text{ft}^3}$$

Continuity equation can be written as :

$$(\rho V A)_{O_2} + (\rho V A)_{H_2} = (\rho V A)_{mix.}$$

$$A_{O_2} = \pi \frac{d^2}{4} = \pi \frac{0.2^2}{4} = A_{H_2} \ , \quad A_{mix.} = \pi \frac{0.4^2}{4}$$

$$\therefore \quad 0.0034 \times 10 \times \pi \frac{0.2^2}{4} + 0.000193 \times 12 \times \pi \frac{0.2^2}{4} = \rho \times 10 \times \pi \frac{0.4^2}{4}$$

$$0.0034 \times 10 + 0.000193 \times 12 = \rho \times 10 \times 4$$

$$\rho = 0.000908 \ \frac{\text{slug}}{\text{ft}^3} = 0.0292 \ \frac{\text{lbm}}{\text{ft}^3}$$

3-12

The velocity in the 4" line is 6 fps and in the 6" line it is 3 fps. Find Q in the 4", 6" and 10" lines. What is the velocity in the 10" line? Determine G (weight flow rate) and the mass flow rate in the 4", 6" and 10" lines. What is V and Q in the 5" line?

$\gamma = 62.4\ \text{Lb/ft}^3$

$\rho = 1.94\ \frac{\text{slugs}}{\text{ft}^3}$

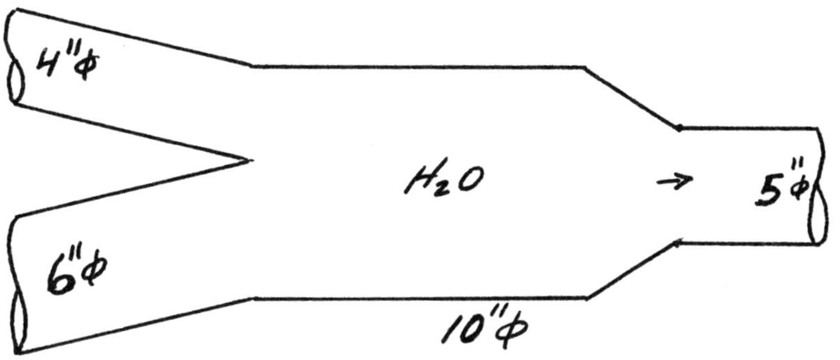

****

$A_4 = \frac{\pi}{4}\left(\frac{4}{12}\right)^2 = .0873\ \text{ft}^2$, $Q_4 = AV = \left(6\ \frac{\text{ft}}{\text{sec}}\right)(.0873\ \text{ft}^2) =$ **.5238 CFS ANS**

$A_6 = \frac{\pi}{4}\left(\frac{6}{12}\right)^2 = .1963\ \text{ft}^2$, $Q_6 = AV = (3)(.1963) =$ **.5889 CFS ANS**

$A_{10} = \frac{\pi}{4}\left(\frac{10}{12}\right)^2 = .5454\ \text{ft}^2$, $Q_{10} = Q_4 + Q_6 =$ **1.1127 CFS ANS**

$A_5 = \frac{\pi}{4}\left(\frac{5}{12}\right)^2 = .1364\ \text{ft}^2$,

$Q = AV;\ V_{10} = \frac{Q}{A} = \frac{1.1127\ \text{CFS}}{.5454\ \text{ft}^2} =$ **2.04 ft/sec ANS**

$G = A\gamma V;\ G_4 = A_4 \gamma V_4 = (.0873\ \text{ft}^2)(62.4\ \text{Lb/ft}^3)(6\ \text{ft/sec}) =$ **32.7 Lb/sec**

$G_6 = A\gamma V = (.1963)(62.4)(3) =$ **36.74 Lb/sec ANS**

$G_{10} = A\gamma V = (.5454)(62.4)(2.04) =$ **69.4 Lb/sec ANS**

$\dot{m}_4 = \rho AV = (1.94)(.0873)(6) =$ **1.02 slug/sec ANS**

$\dot{m}_6 = \rho AV = (1.94)(.1963)(3) =$ **1.14 slug/sec ANS**

$\dot{m}_{10} = \rho AV = (1.94)(.5454)(2.04) =$ **2.15 slug/sec ANS**

$Q_5 = Q_{10} = 1.1127$ CFS ANS

$V_5 = \frac{Q}{A} = \frac{1.1127}{.1364} =$ **8.16 ft/sec ANS**

3-13

You have been hired as a consultant to OWLS Toy Manufacturing Co. to help with the design of a new water pistol. The face of the piston attached to the trigger is 1 in. high and ½ in. wide. Calculate the minimum "barrel" diameter for a fully loaded gun if the trigger can be pulled back at a rate of 1 in. per .5 s. and the muzzle velocity is not to exceed 120 ft./s. for safe play.

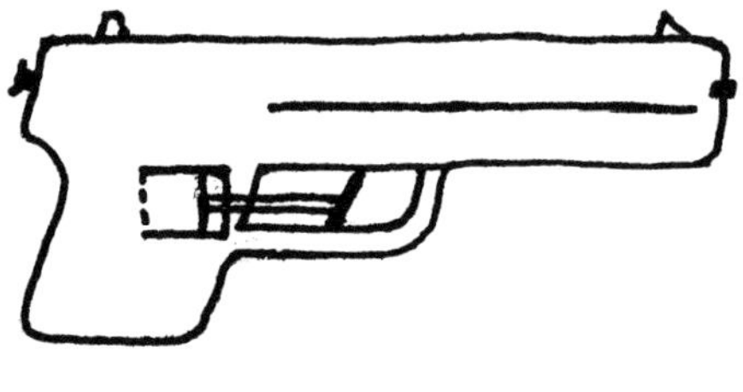

****

Conservation of Mass (Continuity Equation): $\iint_{CS} \rho(\underset{\sim}{v}\cdot\underset{\sim}{n})dA + \frac{\partial}{\partial t}\iiint_{CV} \rho\, dV = 0$ (the time-derivative term is struck through, with an arrow to 0)

cs - control surface cv - control volume

Assume water is an incompressible fluid (This is usually a valid assumption.).

Then,

$$\iint_{CS} (\underset{\sim}{v}\cdot\underset{\sim}{n})dA = 0 \quad \text{or} \quad \iint_{①} (\underset{\sim}{v}\cdot\underset{\sim}{n})dA + \iint_{②} (\underset{\sim}{v}\cdot\underset{\sim}{n})dA = 0$$

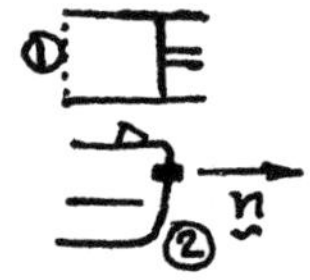

since $(\underset{\sim}{v}\cdot\underset{\sim}{n})$ is zero elsewhere.

Assuming a uniform velocity profile at ① and ②: $-v_1 A_1 + v_2 A_2 = 0$

Hence, $A_2 = \dfrac{\pi d^2}{4} = \dfrac{v_1 A_1}{v_2}$ $\qquad v_1 = 1\text{ in}/.5\text{s} = 2\text{ in/s}$ $\qquad v_2 = 120\text{ ft/s} = 1440\text{ in/s}$

$A_1 = (1\text{ in})(.5\text{ in}) = .5\text{ in}^2$

$$d = \sqrt{\frac{4 v_1 A_1}{\pi v_2}} = 2\sqrt{\frac{(2\text{in/s})(.5\text{in}^2)}{\pi(1440\text{ in/s})}} = .0298\text{ in} = 2.48\cdot10^{-3}\text{ ft}$$

3-14

In a two dimensional flow, velocity components are given by

$$u = A\,y\,(x^2 + y^2) \quad , \quad v = -A\,x\,(x^2 + y^2)$$

Determine whether this flow is possible or not. Is this flow irrotational ?

****

Continuity equation for 2-D case :

$$\frac{\partial u}{\partial x} + \frac{\partial v}{\partial y} = 0$$

$$(2Ayx) + (-2Axy) = 0$$

Continuity is satisfied; this flow is possible.

Vorticity should vanish if the flow is irrotational.

Vorticity $\quad \omega = \dfrac{\partial v}{\partial x} - \dfrac{\partial u}{\partial y}$

$$= (-3Ax^2 - Ay^2) - (Ax^2 + 3Ay^2)$$

$$= -4Ax^2 - 4Ay^2 \neq 0$$

Flow is not irrotational.

BERNOULLI'S EQUATION

3-15

A uniform diameter stream of water falling vertically through the atmosphere has a velocity of 24 meters/sec at an elevation of 12 meters. What is its velocity at an elevation of 4.5 meters?

****

$$\frac{V_1^2}{2g} + Z_1 + \frac{P_1}{\gamma} = \frac{V_2^2}{2g} + Z_2 + \frac{P_2}{\gamma}$$

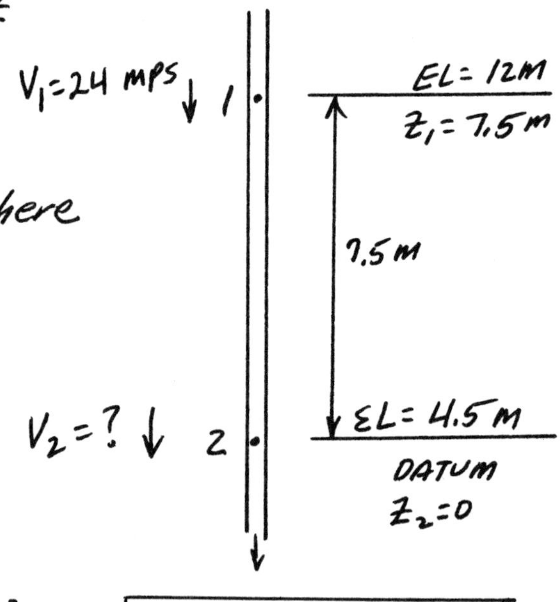

$P_1 = 0$, $P_2 = 0$ } stream is open to the atmosphere

$Z_2 = 0$ (DATUM)

$\therefore$

$$\frac{V_1^2}{2g} + Z_1 = \frac{V_2^2}{2g}$$

$$\frac{(24)^2}{2\,(9.8)} + 7.5 = \frac{V_2^2}{2(9.8)} \; ; \quad \boxed{V_2 = 26.9 \text{ m/s}} \text{ ANS}$$

3-16

The sealed tank has an air pressure of 3000 psfg above the water surface. Water discharges through a 6-inch diameter pipe then a 1-inch diameter pipe. Compute the flow velocity at point A in the 6-inch diameter pipe.

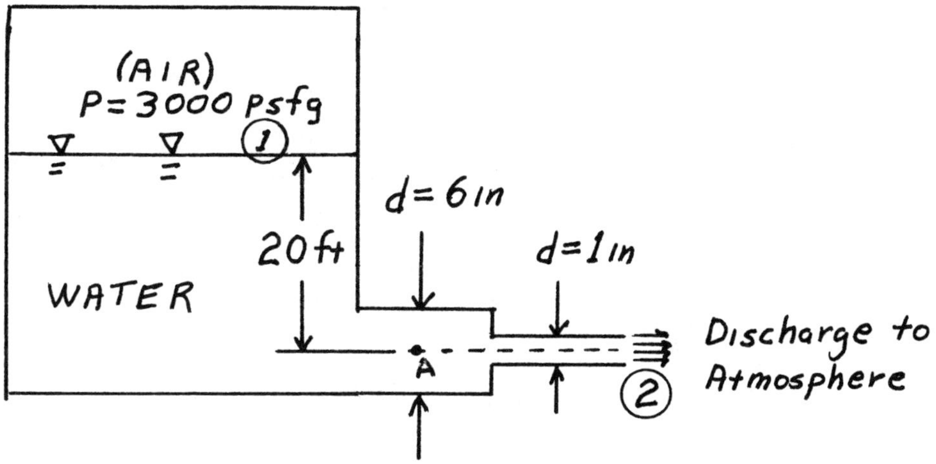

****

Apply Bernoulli's equation between the water surface, point 1, and discharge, point 2.

$$\frac{P_1}{\gamma} + z_1 + \frac{V_1^2}{2g} = \frac{P_2}{\gamma} + z_2 + \frac{V_2^2}{2g}$$

$$\frac{3000\ lb/ft^2}{62.4\ lb/ft^3} + 20\,ft + 0 = 0 + 0 + \frac{V_2^2}{(2)(32.2\ \frac{ft}{s^2})}$$

$$V_2 = \sqrt{(68.1\ ft)(2)(32.2\ ft/s^2)}$$

$$V_2 = 66.2\ ft/sec = \text{Velocity in 1-inch pipe}$$

Apply Continuity equation between the 6-inch and 1-inch pipe: $Q_{6\text{-}in} = Q_{1\text{-}in}$

$$A_{6\text{-}in} V_{6\text{-}in} = A_{1\text{-}in} V_{1\text{-}in}$$

$$V_{6\text{-}in} = (A_{1\text{-}in}/A_{6\text{-}in})\, V_{1\text{-}in}$$

$$V_{6\text{-}in} = (1/36)\ 66.2\frac{ft}{s}$$

$$\underline{\underline{V_{6\text{-}in} = 1.73\ ft/s}}$$

3-17

Water is contained in a tank and is being discharged into the open atmosphere through a 6 in. diameter pipe as shown. Determine the rate of discharge and the gage pressure at the point A. Assume the fluid to be ideal and consider the density of the fluid to be 1.940 $slug/ft^3$.

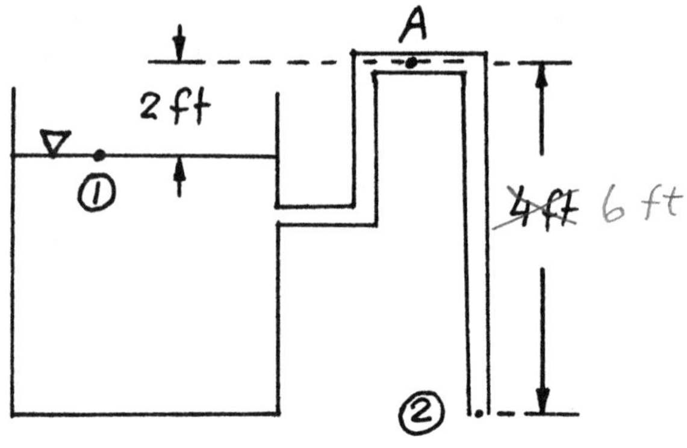

****

(i) Writing Bernoulli Eq. between points 1 and 2, and considering gage pressures only:

$$\frac{p_1}{\rho g} + \frac{V_1^2}{2g} + z_1 = \frac{p_2}{\rho g} + \frac{V_2^2}{2g} + z_2$$

or, $$0 + 0 + 4 = 0 + \frac{V_2^2}{2g} + 0$$

or, $$V_2 = \sqrt{2gz_1} = \sqrt{2 \times 32.174 \times 4} = 16.043 \frac{ft}{sec}$$

(ii) $$Q = A_2 V_2 = \frac{\pi}{4}\left(\frac{1}{4}\right) \times 16.043 = 3.150 \text{ cfs}$$

(iii) Writing Bernoulli Eq. between points 1 and A, and considering gage pressures only:

$$\frac{p_1}{\rho g} + \frac{V_1^2}{2g} + z_1 = \frac{p_A}{\rho g} + \frac{V_A^2}{2g} + z_A$$

or, $$0 + 0 + 0 = \frac{p_A}{\rho g} + 4 + 2$$

or, $$p_A = -6\rho g = -\frac{6 \times 1.940 \times 32.174}{144}$$

or, $$p_A = -2.6 \text{ psi gage}$$

3-18

Water is flowing from an open elevated tank through a hose and nozzle as shown below. The nozzle is fixed at an angle of 35 degrees with the horizontal. Determine the maximum height, h, which the water will reach. Disregard friction.

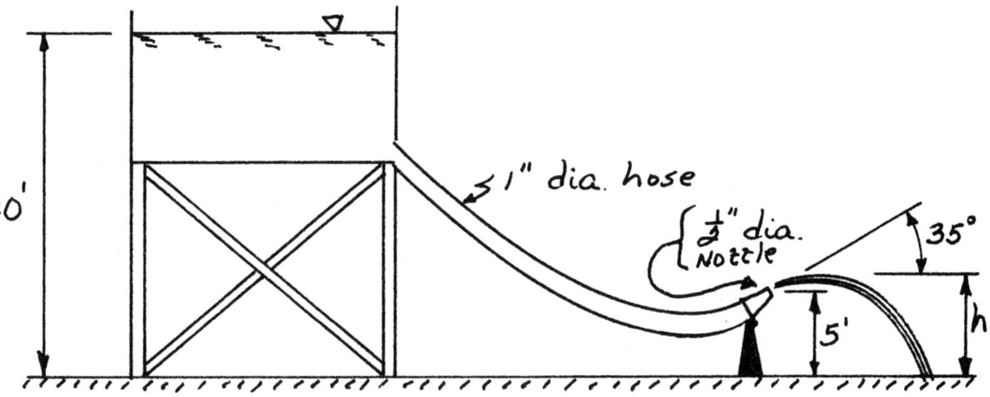

****

let point ① be water surface in the tank
let point ② be the ½-inch dia end of the nozzle
let point ③ be the point of highest discharge at height, h.

$$\frac{V_1^2}{2g} + z_1 + \frac{P_1}{\gamma} = \frac{V_2^2}{2g} + z_2 + \frac{P_2}{\gamma}$$

$$0 + 40\,\text{ft} + 0 = \frac{V_2^2}{2 \times 32.2\,\frac{\text{ft}}{\text{sec}^2}} + 5\,\text{ft} + 0 \Rightarrow V_2 = 47.5\,\frac{\text{ft}}{\text{sec}}$$

V_3 = horiz component of $V_2 = 47.5\,\frac{\text{ft}}{\text{sec}} \cos 35 = 38.9\,\frac{\text{ft}}{\text{sec}}$

(NOTE: $V_3 = 38.9$ fps is true if air resistance is neglected. V_3 also has no vertical comp.)

$$\frac{V_1^2}{2g} + z_1 + \frac{P_1}{\gamma} = \frac{V_3^2}{2g} + z_3 + \frac{P_3}{\gamma}$$

$$0 + 40' + 0 = \frac{(38.9\,\text{ft/sec})^2}{2 \times 32.2\,\text{ft/sec}^2} + h + 0 \Rightarrow \underline{\underline{h = 16.5\,\text{ft}}}$$

3-19

A water tank is being filled with water from a large reservoir as shown below. The diameter of the water tank is 5 ft, and the water level is initially 6 ft. It is desired to fill the tank to a level of 30 ft.

a. How long after the jet from the reservoir is turned on before the level in the tank is 30 ft?
b. Assume now that a leak is discovered 3 ft above the bottom of the tank. The leak may be modeled as a circular hole of 6" diameter.
   (i) Can the steady form of Bernoulli's equation be used to calculate the velocity of the water exiting the leak as the tank is being filled? Why or why not.
   (ii) Assume that the steady Bernoulli equation may be used to calculate the velocity of the water exiting at the leak. What is the time required to fill the tank to a level of 30 ft?
   (iii) If the jet is allowed to run indefinitely what is the limit of the water level in the tank as $t \to \infty$?

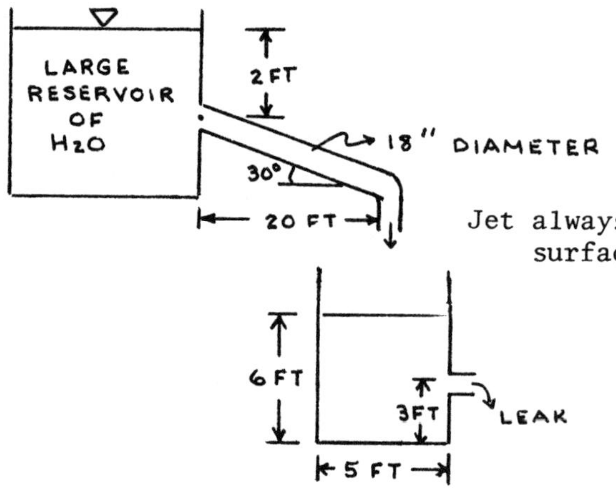

Jet always remains above the surface of water.

****

a) THE VOLUME OF WATER TO BE FILLED IN THE TANK IS

$$V = A\,\Delta h = \frac{\pi d^2}{4}\Delta h = \frac{\pi (5\text{ FT})^2}{4}(30\text{ FT} - 6\text{ FT}) = 471.24\text{ FT}^3$$

THE FLOW RATE INTO THE TANK CAN BE CALCULATED BY APPLYING BERNOULLI'S EQUATION BETWEEN THE SURFACE OF THE RESEVOIR (1) AND THE PIPE OUTLET (2) ASSUMING NO FRICTIONAL LOSSES

$$P_1^{\,0} + \frac{1}{2}\rho V_1^{2\,0} + \rho g z_1 = P_2^{\,0} + \frac{1}{2}\rho V_2^2 + \rho g z_2$$

$$V_2^2 = 2g(z_1 - z_2) \quad \text{with } z_2 \text{ as datum}$$

$$V_2^2 = [2(32.2\text{ FT/sec}^2)(2\text{ FT} + 20\text{ FT}\tan 30^\circ)]$$

$$V_2 = 29.54\text{ FT/SEC}$$

THE FLOW RATE INTO THE TANK IS

$$Q_2 = A_2 V_2 = \pi \left(\frac{18 \text{ IN}}{2 (12 \text{ IN/FT})} \right)^2 29.54 \text{ FT/SEC}$$

$$= 52.20 \text{ FT}^3/\text{SEC}$$

IF THE FLOW IS STEADY THEN THE TIME REQUIRED TO FILL THE TANK IS

$$t = V/Q_2 = \frac{471.2 \text{ FT}^3}{52.20 \text{ FT}^3/\text{SEC}} = 9.03 \text{ SEC}$$

6) i) NO, THE VELOCITY OF THE LEAK IS CHANGING WITH TIME SINCE THE HEIGHT OF THE LIQUID IN THE TANK CHANGING WITH TIME. AN UNSTEADY FORM OF BERNOULLI'S EQUATION SHOULD BE USED.

ii) $\frac{dV}{dt} = Q_{in} - Q_{out}$ → CONSERVATION OF MASS OF FLUID IN THE TANK

$Q_{IN} = Q_2 = 52.2 \text{ FT}^3/\text{SEC}$

BERNOULLI'S EQUATION BETWEEN (3) THE SURFACE OF THE TANK AND (4) THE OUTLET OF THE LEAK

$$\cancel{P_3}^{\,0} + \tfrac{1}{2}\rho \cancel{V_3^2}^{\,0 \text{ (ASSUMED SMALL)}} + \rho g h_3 = \cancel{P_4}^{\,0} + \tfrac{1}{2}\rho V_4^2 + \rho g h_4$$

$$V_4 = \sqrt{2gh}$$

$$Q_{OUT} = A_{hole} V_4 = A_{hole}\sqrt{2gh}$$

WHERE h IS THE FLUID LEVEL ABOVE THE LEAK

$$\frac{dV}{dt} = \frac{d}{dt}(A_{tank}\, h) = A_{tank}\frac{dh}{dt}$$

THUS

$$A_{tank}\frac{dh}{dt} = 52.2 - A_{hole}\sqrt{2gh}$$

SEPARATING VARIABLES

$$dt = A_{tank}\frac{dh}{52.2 - A_{hole}\sqrt{2gh}}$$

WHEN $t=0$, $h = 3$ FT. WE DESIRE TO FIND THE VALUE $\tilde{t}$ SUCH THAT $h = 27$ FT. THUS WE INTEGRATE THE ABOVE

EQUATION BETWEEN THESE LIMITS

$$\int_0^{\tilde{t}} dt = \int_{3\,FT}^{27\,FT} \frac{A_{tank}\, dh}{52.2 - A_{hole}\sqrt{2gh}}$$

$$\tilde{t} = \int_{3\,FT}^{27\,FT} \frac{A_{tank}\, dh}{52.2 - A_{hole}\sqrt{2gh}}$$

THE INTEGRAL CAN BE EVALUATED THROUGH A SIMPLE CHANGE OF VARIABLES

$$u = A_{hole}\sqrt{2gh} \qquad h^{1/2} = \frac{u}{A_{hole}\sqrt{2g}}$$

$$du = A_{hole}\sqrt{2g}\ \tfrac{1}{2} h^{-1/2} dh = \frac{A_{hole}^2 (2g)}{2u}\, dh$$

$$dh = \frac{2u}{A_{hole}^2 2g}\, du$$

THUS

$$\int \frac{dh}{52.2 - A_{hole}\sqrt{2gh}} = \int \frac{2u\, du}{A_{hole}^2 (2g)(52.2 - u)}$$

$$= \frac{1}{A_{hole}^2 g} \int \left(-1 + \frac{52.2}{u}\right) du$$

$$= \frac{1}{A_{hole}^2 g} \left(-u + 52.2 \ln u\right)$$

THUS

$$\tilde{t} = \frac{A_{tank}}{A_{hole}^2 g} \left[-A_{hole}\sqrt{2gh} + 52.2 \ln A_{hole}\sqrt{2gh} \right]_{h = 3\,FT}^{h = 27\,FT} = 10.62\ SEC$$

iii) A STEADY STATE WILL EXIST WHEN THE FLOW RATE IN EQUALS THE FLOW RATE OUT.

$$Q_{IN} - Q_{OUT} = 0$$

OR

$$A_{hole}\sqrt{2gh} = 52.2\ FT^3/SEC$$

$$h = 1097.5\ FT$$

3-20

Water is flowing from a reservoir through a pipe with a constriction and discharges into the open. Assuming ideal flow, calculate the flow rate that can be obtained. The atmospheric pressure is 101.3 kPa and the water temperature is 10°C.

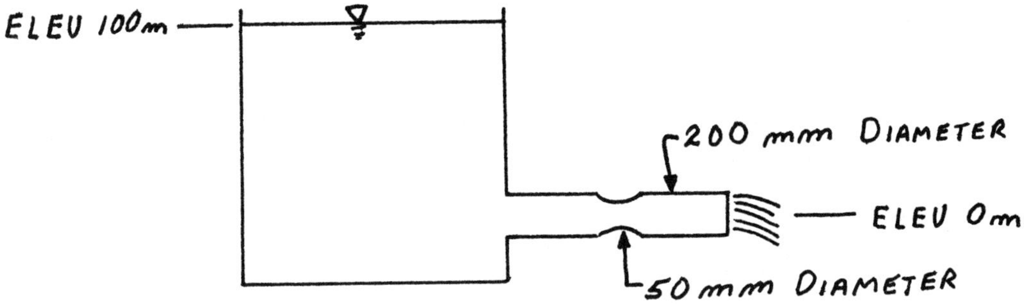

****

MAXIMUM FREE FLOW WITHOUT CONSTRICTION IS

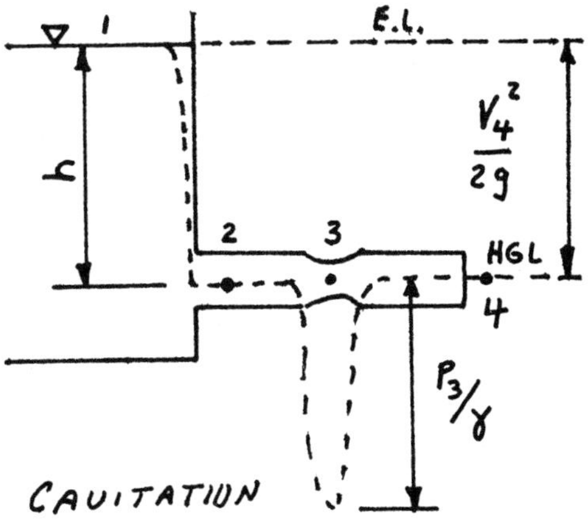

$$Q = AV = A\sqrt{2gh} = \frac{\pi d^2}{4}\sqrt{2gh}$$

$$= (3.14)\frac{(.2)^2}{4}\sqrt{(2)(9.81)(100)}$$

$$= \underline{\underline{1.39\ m^3/s}}$$

SOLVE FOR MAXIMUM FLOW WITH CAVITATION JUST STARTING AT POINT 3.

$$\frac{P_3}{\gamma} + \frac{V_3^2}{2g} + z_3 = h \qquad \frac{P_V - P_{ATM}}{\gamma_{H_2O}} + \frac{V^3}{2g} + z_3 = h$$

$$\frac{1.23\ kPa - 101.3\ kPa}{9.804\ kN/m^3} + \frac{V_3^2}{(2)(9.81\ m/s^2)} + 0\ m = 100\ m$$

$$V_3 = 46.5\ m/s$$

$$Q = AV = \frac{\pi d^2}{4}V = \frac{(3.14)(.05)^2}{4}(46.5) = \underline{\underline{0.091\ m^3/s}}$$

3-21

An airfoil moves in stagnant air with a velocity of 80 ft/s. Due to the curvature of the upper and lower surfaces of the airfoil, air molecules are accelerated 1.3 and 1.05 times on the upper and lower surfaces of this airfoil, respectively. Air is at standard atmospheric conditions away from the airfoil (14.7 psia and 80°F). Assume incompressible flow around the airfoil. Determine the maximum pressure difference between the upper and lower surfaces of this airfoil.

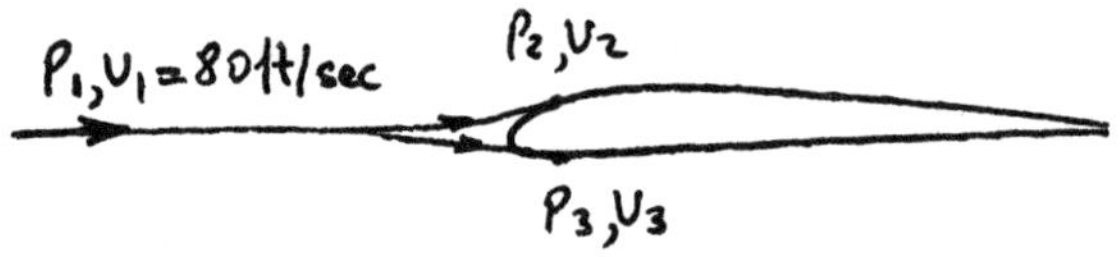

****

Density of air : $\rho = \dfrac{P}{RT} = \dfrac{14.7 \times (144) \times 28.97}{1545.3 \times (80+460)} = 0.0735 \dfrac{lbm}{ft^3}$

$= 0.002284 \dfrac{slug}{ft^3}$

For 1-2 streamline, at point 2, $V_2 = 80 \times 1.3 = 104$ ft/sec

$$P_1 + \frac{1}{2}\rho V_1^2 = P_2 + \frac{1}{2}\rho V_2^2 \longrightarrow P_2 = P_1 + \frac{1}{2}\rho\,(V_1^2 - V_2^2)$$

For 1-3 streamline, at point 3, $V_3 = 80 \times 1.05 = 84$ ft/sec

$$P_1 + \frac{1}{2}\rho V_1^2 = P_3 + \frac{1}{2}\rho V_3^2 \longrightarrow P_3 = P_1 + \frac{1}{2}\rho\,(V_1^2 - V_3^2)$$

Therefore, pressure difference $P_3 - P_2 = \frac{1}{2}\rho\,(V_2^2 - V_3^2)$

$$= \frac{1}{2} \times 0.002284\,(104^2 - 84^2)$$

$$P_3 - P_2 = 4.29\ lbf/ft^2$$

4 MOMENTUM PRINCIPLE

FORCES IN PIPES

4-1

Oil (s.g. = 0.86) is flowing through the horizontal pipeline shown below at the rate of 10,000 gpm. The gage reading is 10 psi. Determine the horizontal force on the contraction.

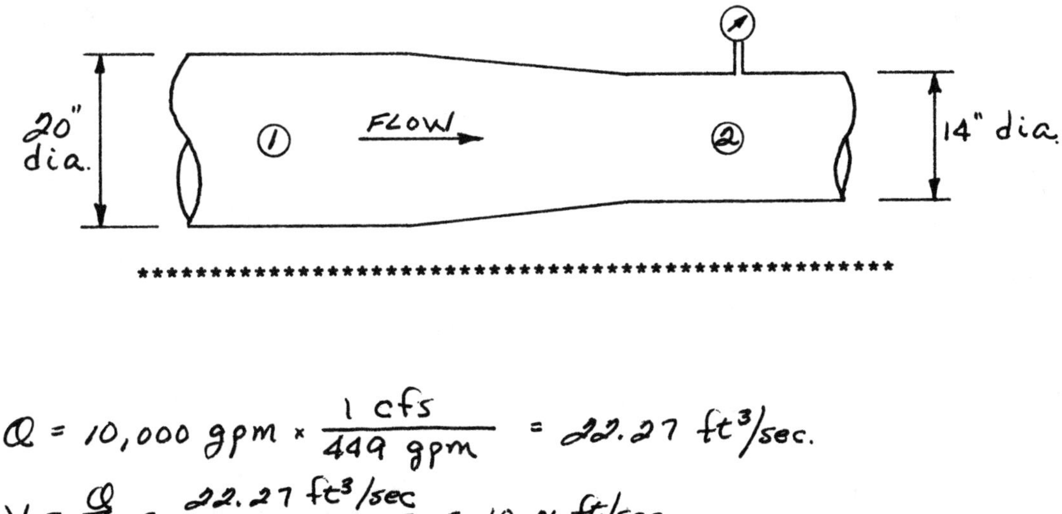

****

$$Q = 10{,}000 \text{ gpm} \times \frac{1 \text{ cfs}}{449 \text{ gpm}} = 22.27 \text{ ft}^3/\text{sec.}$$

$$V_1 = \frac{Q}{A_1} = \frac{22.27 \text{ ft}^3/\text{sec}}{\frac{\pi}{4}\left(\frac{20}{12}\text{ ft}\right)^2} = 10.21 \text{ ft/sec.}$$

$$V_2 = \frac{Q}{A_2} = \frac{22.27 \text{ ft}^3/\text{sec}}{\frac{\pi}{4}\left(\frac{14}{12}\text{ ft}\right)^2} = 20.83 \text{ ft/sec}$$

$$\frac{V_1^2}{2g} + z_1 + \frac{P_1}{\gamma} = \frac{V_2^2}{2g} + z_2 + \frac{P_2}{\gamma}$$

since pipe is horizontal, $z_1 = z_2$

$$\therefore \frac{p_1}{\gamma} = \frac{V_2^2 - V_1^2}{2g} + \frac{p_2}{\gamma}$$

$$\frac{p_1}{\gamma} = \frac{(20.83\ \text{ft/sec})^2 - (10.21\ \text{ft/sec})^2}{2 \times 32.2\ \text{ft/sec}^2} + \frac{10\ \frac{\text{lb}}{\text{IN}^2} \times \frac{144\ \text{IN}^2}{\text{ft}^2}}{0.86 \times 62.4\ \text{lb/ft}^3}$$

$$\frac{p_1}{\gamma} = 31.95\ \text{ft} \Rightarrow p_1 = 31.95\ \text{ft} \times 0.86 \times 62.4\ \frac{\text{lb}}{\text{ft}^3} \times \frac{1\ \text{ft}^2}{144\ \text{IN}^2} = 11.91\ \text{psi}$$

F.B.D. of oil in contraction:

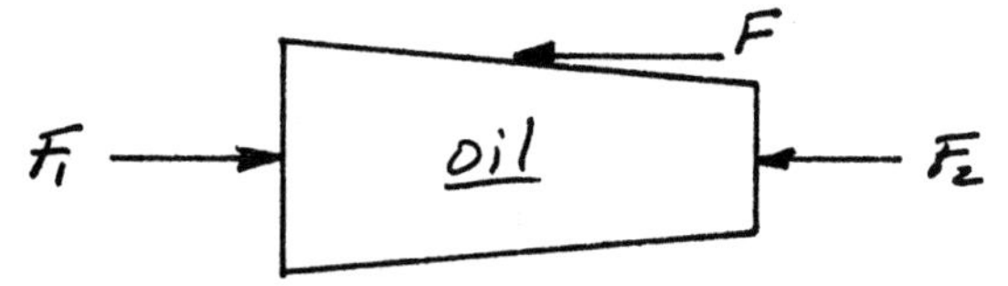

F_1 = force due to $p_1 = p_1 A_1 = 11.91\ \frac{\text{lb}}{\text{IN}^2} \times \frac{\pi}{4}(20\ \text{in})^2 = 3741.6\ \text{lb.}$

F_2 = force due to $p_2 = p_2 A_2 = 10\ \frac{\text{lb}}{\text{IN}^2} \times \frac{\pi}{4}(14\ \text{in.})^2 = 1539.4\ \text{lb.}$

F = horiz component of force exerted by the contraction on the oil.

$\overset{+}{\rightarrow}$ $\Sigma F_x = Q\rho(V_2 - V_1)_x \Rightarrow F_1 - F_2 - F = Q\rho(V_2 - V_1)_x$

$$\therefore F = F_1 - F_2 - Q\rho(V_2 - V_1)_x$$

$$F = 3741.6\ \text{lb} - 1539.4\ \text{lb} - \left(22.27\ \frac{\text{ft}^3}{\text{sec}}\right)\left(\frac{.86 \times 62.4\ \text{lb/ft}^3}{32.2\ \text{ft/sec}^2}\right) \times (20.83 - 10.21)\ \text{ft/sec}$$

$$F = (3741.6 - 1539.4 - 394.2)\ \text{lb} = \overleftarrow{1808\ \text{lb}} = \text{force on the oil}$$

$\therefore$ horiz force on contraction $= \overrightarrow{1808\ \text{lb}}$

4-2

A four inch diameter water pipe terminates at a tee. With a flowrate of 1.0 cfs determine the magnitude of the force that must be applied at the tee to resist the resultant force of the water on the tee.

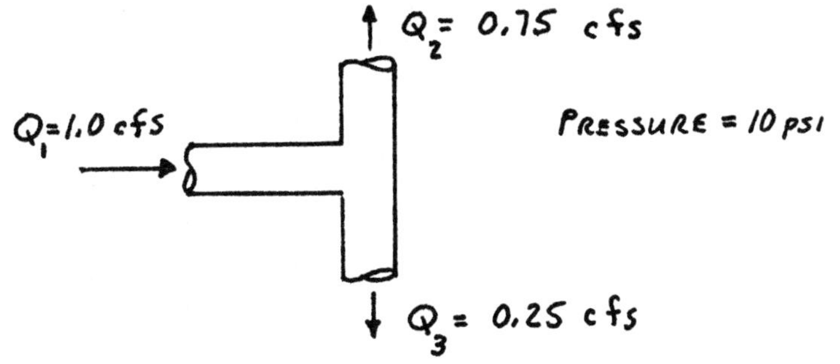

****

SOLVE FOR UNKNOWN VALUES

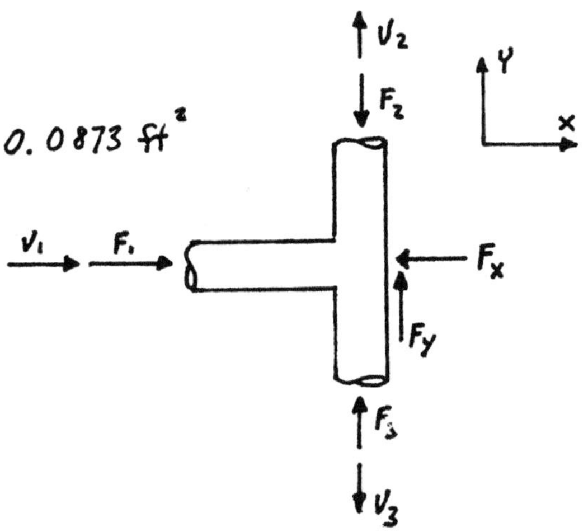

$$A_1 = A_2 = A_3 = \frac{\pi d^2}{4} = \frac{(3.14)(4/12)^2}{4} = 0.0873 \text{ ft}^2$$

$$V_1 = \frac{Q_1}{A} = \frac{1.0 \text{ cfs}}{0.0873 \text{ ft}^3} = 11.45 \text{ fps}$$

$$V_2 = \frac{Q_2}{A} = \frac{0.75}{0.0873} = 8.59 \text{ fps}$$

$$V_3 = \frac{Q_3}{A} = \frac{0.25}{0.0873} = 2.86 \text{ fps}$$

$$F_1 = F_2 = F_3 = PA = (10 \text{ psi})(144)(0.0873) = 125.7 \text{ LB}$$

SOLVE FOR IMPULSE-MOMENT FORCES USING $\Sigma F = Q\rho(V_o - V_i)$

$$\Sigma F_x = F_1 - F_x = Q\rho(V_{OUT} - V_{IN}) \qquad \Sigma F_x = 125.7 - F_x = (1 \text{ cfs})(1.937)(0 - 11.45)$$

$$F_x = \underline{\underline{147.9 \text{ LB}}}$$

$$\Sigma F_y = F_3 - F_2 + F_y = (Q\rho V)_{OUT} - (Q\rho V)_{IN} = Q_2 \rho V_2 + Q_3 \rho V_3 - 0$$

$$125.7 - 125.7 + F_y = (0.75)(1.937)(8.59) + (0.25)(1.937)(-2.86) - 0$$

$$F_y = \underline{\underline{11.1 \text{ LB}}} \qquad F = \sqrt{(147.9)^2 + (11.1)^2} = \underline{\underline{148.4 \text{ LB}}}$$

$$\phi = \text{TAN}^{-1}\frac{F_y}{F_x} = \text{TAN}^{-1}\frac{11.1}{147.9} = \underline{\underline{4.29^\circ}}$$

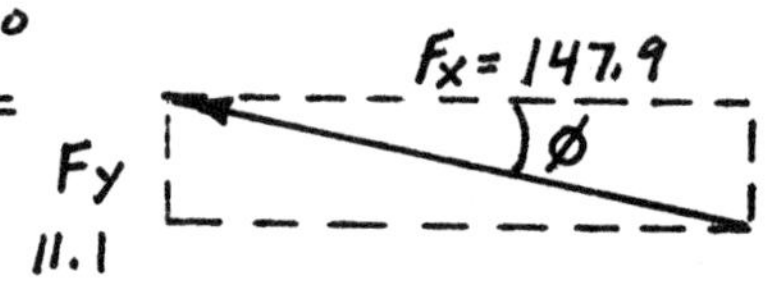

$F = 148.4$ LB

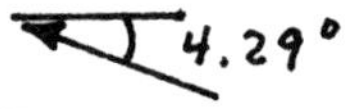

4-3

A pipe branching system is shown below. Water flows into the system at A with a velocity of 10 m/sec at a pressure of 10 kPa. Water exits into the atmosphere at B and C. The velocity of the water leaving at B is measured as 20 m/sec. The system is in a horizontal plane and is supported at D. Calculate the reactions of the pipe on its support. List all assumptions.

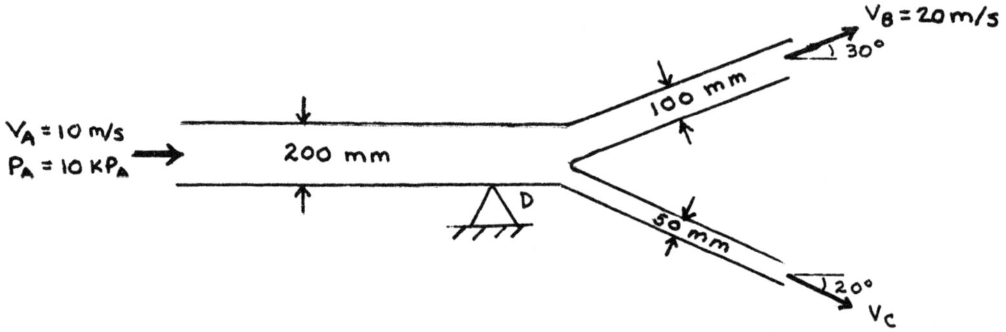

****

ASSUMPTIONS

1. STEADY
2. INCOMPRESSIBLE
3. INVISCID
4. UNIFORM FLOW AT A, B & C

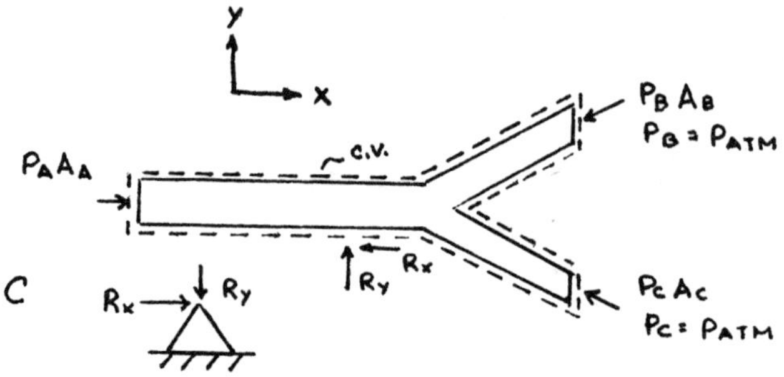

CONSERVATION OF LINEAR MOMENTUM FOR A CONTROL VOLUME

$$\Sigma \vec{F} = \frac{\partial}{\partial t}\int_{c.v.} \rho \vec{V}\, dV \ (\to 0 \text{ (STEADY)}) + \int_{c.s.} \rho \vec{V}\, \vec{V}\cdot\vec{n}\, d\sigma$$

OR

$$P_A A_A \vec{i} + R_y \vec{j} - R_x \vec{i} = \rho\big[V_A \vec{i}\,(V_A \vec{i} \cdot -\vec{i})\, A_A$$
$$+ V_B(\cos 30^\circ \vec{i} + \sin 30^\circ \vec{j}) V_B (\cos 30^\circ \vec{i} + \sin 30^\circ \vec{j}) \cdot A_B$$
$$+ V_C(\cos 20^\circ \vec{i} - \sin 20^\circ \vec{j}) V_C (\cos 20^\circ \vec{i} - \sin 20^\circ \vec{j}) \cdot A_C \big]$$

$\vec{i}$ COMPONENTS

$$R_x = P_A A_A + \rho V_A^2 A_A - \rho V_B^2 \cos 30^\circ A_B - \rho V_C^2 \cos 20^\circ A_C$$

$\vec{j}$ COMPONENTS

$$R_y = \rho V_B^2 \sin 30^\circ A_B - \rho V_C^2 \sin 20^\circ A_C$$

CONTINUITY: $Q_A = Q_B + Q_C$

$$V_A A_A = V_B A_B + V_C A_C \longrightarrow V_C = 80 \text{ m/sec}$$

$$\therefore R_x = (10000 \text{ N/m}^2)(0.1\text{ m})^2\pi + 1000 \tfrac{\text{Kg}}{\text{m}^3}\big[(10 \text{ m/sec})^2(0.1\text{ m})^2\pi$$
$$- (20 \text{ m/sec})^2 (0.05\text{ m})^2 \pi \sqrt{3}/2 - (80 \text{ m/sec})^2 (0.940)(0.025)^2\pi\big]$$
$$= -11{,}080 \text{ N}$$

$$R_y = 1000 \tfrac{\text{kg}}{\text{m}^3}\Big[\big(20 \tfrac{\text{m}}{\text{sec}}\big)^2 \tfrac{1}{2}(0.05\text{m})^2\pi - \big(80 \tfrac{\text{m}}{\text{sec}}\big)^2 (0.342)(0.025)^2\pi\Big]$$
$$= -2730 \text{ N}$$

4-4

Water is flowing at a rate of 2.2 cfs. What is the force on the reducer? Be explicit about the direction of the force.
Ignore head loss through the reducer.
Draw a clear free body diagram of the water.

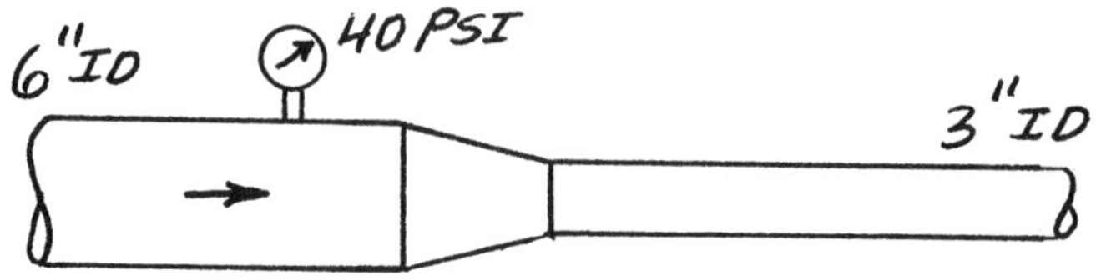

****

$$A_1 = \frac{\pi\left(\frac{6}{12}\right)^2}{4} = .1963 \text{ ft}^2; \quad V_1 = \frac{Q}{A} = \frac{2.2}{.1963} = 11.2 \text{ ft/s}$$

$$A_2 = \frac{\pi}{4}\left(\frac{3}{12}\right)^2 = .0491 \text{ ft}^2 \quad V_2 = \frac{2.2}{.0491} = 44.8 \text{ ft/s}$$

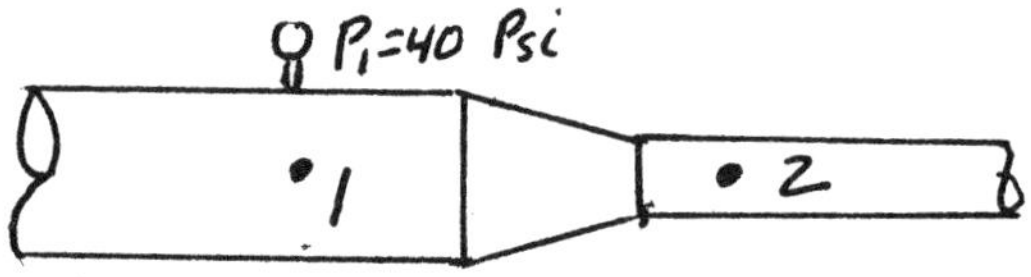

FIND P_2

$$\frac{P_1}{\gamma} + \cancel{z_1} + \frac{V_1^2}{2g} = \frac{P_2}{\gamma} + \cancel{z_2} + \frac{V_2^2}{2g}$$

$$\frac{40(144)}{62.4} + \frac{(11.2)^2}{2(32.2)} = \frac{P_2(144)}{62.4} + \frac{(44.8)^2}{2(32.2)}; \quad P_2 = 27.3 \text{ psi}$$

FBD:

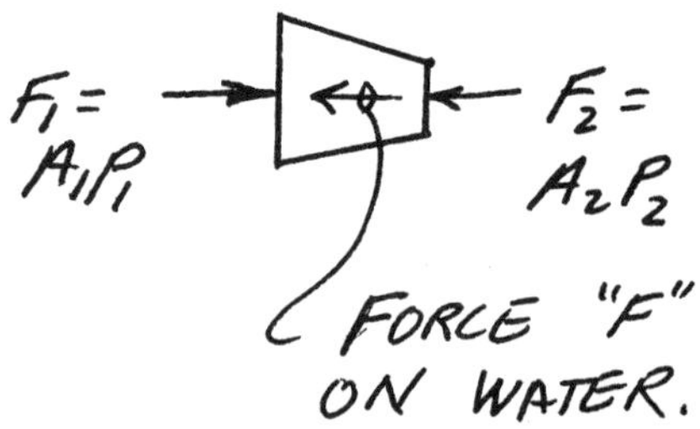

$$F_1 = A_1P_1 = (.1963 \text{ ft}^2)(40 \times 144) = 1131 \text{ lbs}$$

$$F_2 = A_2P_2 = (.0491 \text{ ft}^2)(27.3 \times 144) = 193 \text{ lbs}$$

$$\overset{+}{\rightarrow} \quad \Sigma F_x = \rho Q (V_2 - V_1)$$

$$1131 - F - 193 = 1.936(2.2)(44.8 - 11.2)$$

$F = 795^{\#} \leftarrow$ ON WATER, AS SHOWN IN FBD

$F = 795^{\#} \longrightarrow$ ON REDUCER	ANS

4-5

Calculate the forces to be exerted on a pipe elbow (forces on the elbow) to hold it in equilibrium. The elbow is 4 inch schedule 40 size carrying 3500 L/Min of water at 15°C. The inlet pressure is 500 KPa. Show on the diagram, the forces (x), (y) and resultant. Give an angle locating the resultant.

****

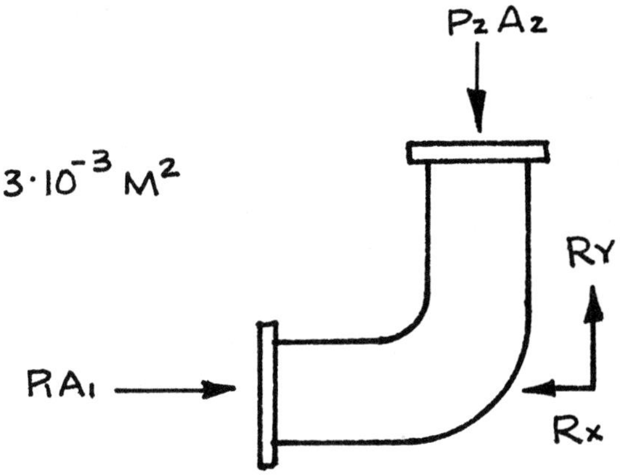

4 IN SCHEDULE 40 $A_i = 8.213 \cdot 10^{-3}\ M^2$

$3500\ L/M = .05833\ M^3/s$

@ 15°C $\rho = 10^3\ Kg/M^3$

$\Sigma F_x = M a_x$ (FORCES ON THE FLUID) $(\longrightarrow +)$

$-R_x + P_1 A_1 = \rho Q (V_{2x} - V_{1x})$ $\Delta V_x = a_x$, $V_{2x} = 0$

$$R_x - 4106.5\ Nt = 10^3 (.05833) \left(\frac{.05833}{8.213 \cdot 10^{-3}} \right)$$

$R_x = 4520.7$ N

DUE TO SYMMETRY $R_Y = 4520.7$ N

RESULTANT ON THE FLUID $R = \sqrt{R_x^2 + R_Y^2}$

$R = 6393$ N

RESULTANT ON THE BEND IS OPPOSITE TO (R) AS SHOWN

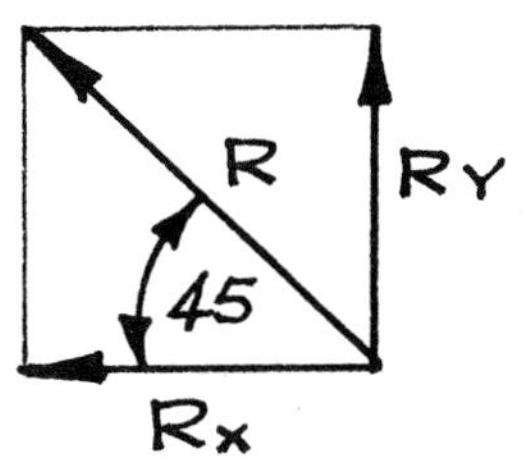

4-6

In the figure shown below, determine the reaction if water at 70°F flows in this pipe.

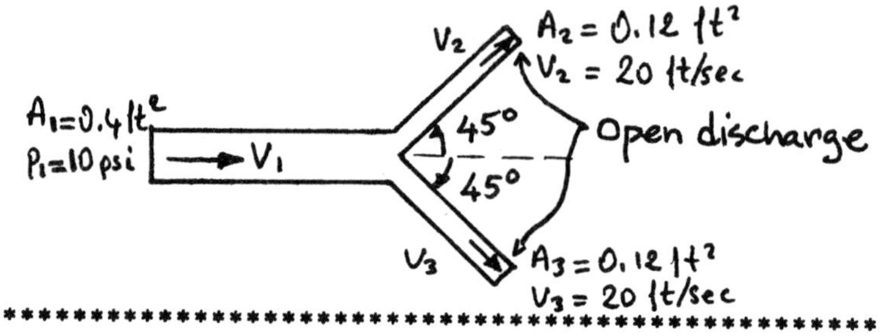

****

Equation of continuity (conservation of mass) : $\rightarrow (+x)$ direction.

$$\rho_1 V_1 A_1 = \rho_2 V_2 A_2 + \rho_3 V_3 A_3 \qquad \rho_1 = \rho_2 = \rho_3 = \rho \text{ incompressible.}$$

$$\therefore \quad 0.4\, V_1 = 0.12 \times 20 + 0.12 \times 20 \rightarrow V_1 = \frac{2 \times 20 \times 0.12}{0.4} = 12 \text{ ft/sec.}$$

Equation of momentum : in the x direction (in the y direction, forces are balanced) :

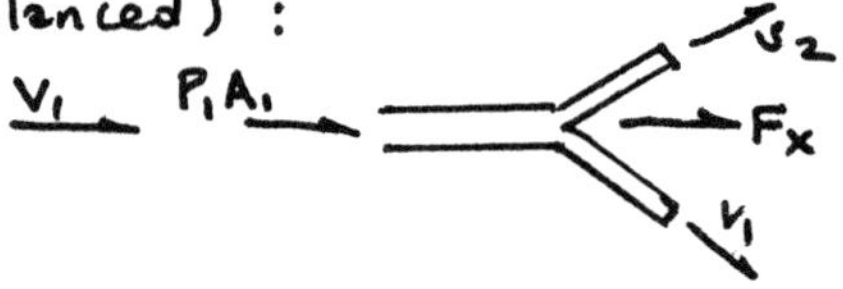

$$\Sigma F_x = \sum \rho V A . V_x$$

$$F_x + P_1 A_1 = (\rho_2 V_2 A_2)\, V_2 \cos 45° + (\rho_3 V_3 A_3)\, V_3 \cos 45° - (\rho_1 V_1 A_1)\, V_1$$

$$F_x = \left(1.94 \times 20 \times 0.12 \times 20 \times \frac{\sqrt{2}}{2}\right) + \left(1.94 \times 20 \times 0.12 \times 20 \times \frac{\sqrt{2}}{2}\right) - (1.94 \times 12 \times 0.4 \times 12) - (10 \times 144 \times 0.4)$$

$F_x = -\ 556$ lbf $\rightarrow$

$= +\ 556$ lbf $\leftarrow$ on water

$=\ 556$ lbf $\rightarrow$ on pipe

4-7

The 90° elbow in the Figure is designed to carry 40,000GPM of water (62.4 lb/ft^3). The joint is secured by ten bolts. What is the tension in each bolt due to the water flow and the pressure P of 20 psig if the joint is designed to take the full impact.

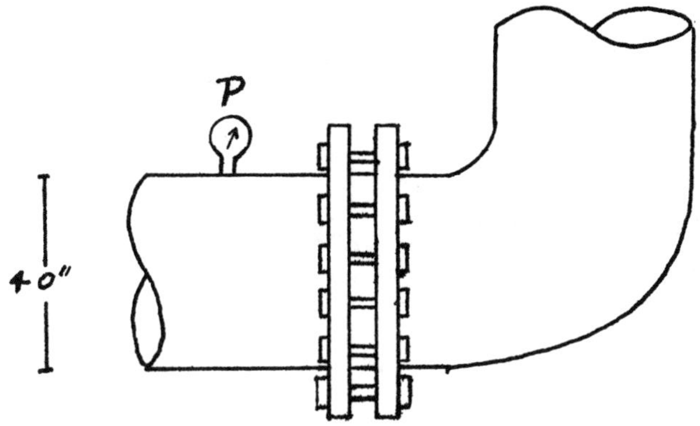

****

The only forces producing tension are horizontal forces.

$R = F_v + F_p$ — R = reacting force

$F_v = \frac{\gamma}{g} Q v$ — F_v = F due to velocity

$F_p = p \times A$ — F_p = F due to pressure

$$Q = 40000\frac{gal}{min} \times \frac{1\,min}{60\,s} \times \frac{1\,ft^3}{7.48\,gal} = 89.12\,\frac{ft^3}{s}$$

$$A = \frac{\pi (40/12)^2}{4} = 8.73\,ft^2$$

$$p = 20\frac{lb}{in^2} \times \frac{144\,in^2}{1\,ft^2} = 2{,}880\,\frac{lb}{ft^2}$$

$$v = \frac{89.12}{8.73} = 10.21\,ft/s$$

$$R = \frac{62.4\,lb/ft^3}{32.2\,ft/s^2} \times 89.12\,ft^3/s \times 10.21\,ft/s + 2{,}880\frac{lb}{ft^2} \times 8.73\,ft^2$$

$$= 26{,}906\,lb$$

for each bolt: 2,691 lb

4-8

2 cfs of water is pumped through the 45° vertical pipe bend in the 6 inch diameter pipeline shown below. Determine the horizontal force (F_x) acting on the thrust block.

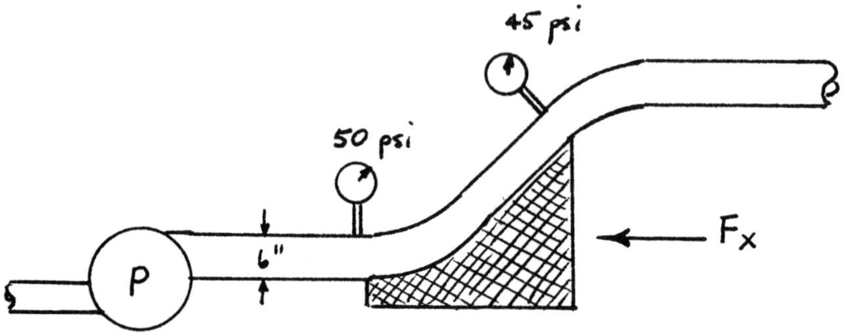

****

Control Volume of 45° bend:

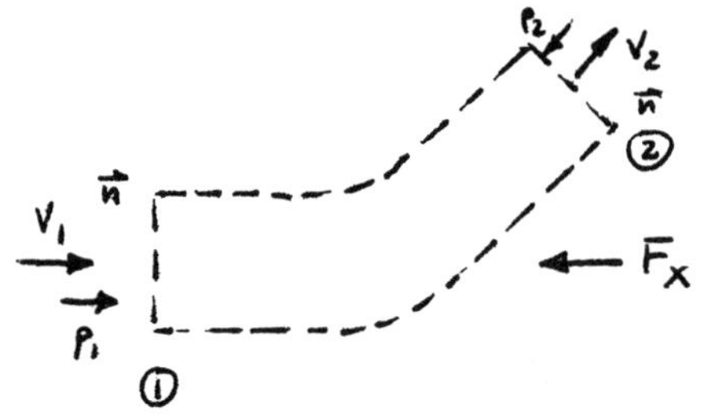

$$A_1 = A_2 = \frac{\pi(6)^2}{4} = 28.3 \text{ in}^2$$

$$V_1 = V_2 = \frac{Q}{A} = \frac{2}{\frac{\pi(.5)^2}{4}}$$

$$V = 10.2 \text{ fps}$$

From Momentum balance:

$$\Sigma F_x = \int_{c.s.} \rho \vec{V}\, \vec{V} \cdot \vec{n}\, dA$$

$$p_1 A_1 - p_2 A_2 \cos 45^\circ - F_x = \rho V_1 (-V_1 A_1) + \rho V_2 \cos 45^\circ (+V_2 A_2)$$

$$(p_1 - p_2 \cos 45^\circ) A - F_x = \rho Q V (\cos 45^\circ - 1)$$

$$(50 - 45 \cos 45^\circ)\, 28.3 - F_x = 1.94(2)(10.2)[.707 - 1]$$

$$514 - F_x = -11.6$$

$$F_x = 526 \text{ lbs} \leftarrow$$

LINEAR AND ANGULAR MOMENTUM

4-9

A water jet with a velocity of 200 ft/sec enters a moving blade as shown below. The absolute velocity of the jet is deflected through an angle of 125 degrees. Determine the blade angle, β, and the magnitude of the final absolute velocity of the jet.

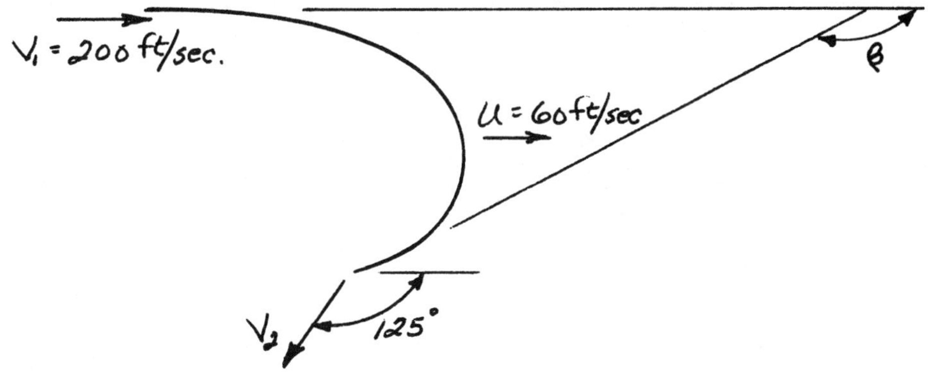

****

initial relative velocity = final relative velocity

$$v_1 = v_2 \Rightarrow v_1 = V_1 - u = 200 - 60 = 140 \text{ ft/sec.} = v_2$$

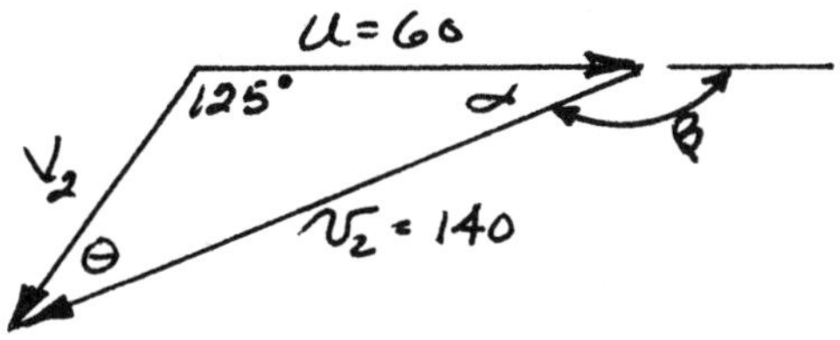

$$\frac{\sin 125}{140} = \frac{\sin\theta}{60}$$

$$\theta = 20.55^\circ$$

$$\alpha = 180 - 125 - 20.55 = 34.45^\circ$$

$$\beta = 180 - 34.45 \Rightarrow \underline{\underline{\beta = 145.55^\circ}}$$

$$\frac{V_2}{\sin 34.45^\circ} = \frac{140}{\sin 125^\circ} \Rightarrow \underline{\underline{V_2 = 96.7 \text{ ft/sec}}}$$

4-10

Air at 12 psia and 60°F enters the jet engine of an aircraft flying at a velocity of 500 ft/s. Inlet area of the engine is 4 ft^2. Exhaust gases leave the engine with a relative velocity of 900 ft/s. Fuel flow rate is only 2% of the mass flow rate of the air. Calculate the thrust force developed by this engine. Also, determine the thrust horsepower.

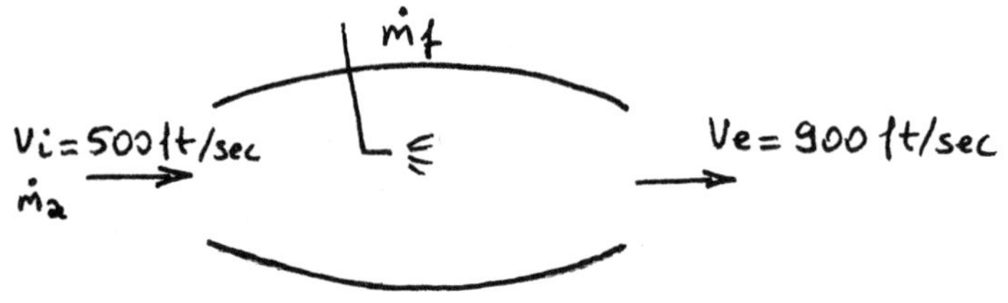

****

Mass flow rate of air $\dot{m}_a = \rho_i V_i A_i$

$$\rho_i = \frac{P}{RT} = \frac{12 \times 144 \times 28.97}{1545.3 \times (460+60)} = 0.0623 \ lb_m/ft^3 = 0.00194 \ slug/ft^3$$

$$\dot{m}_a = 0.00194 \times 500 \times 4 = 3.873 \ slug/sec$$

$$\dot{m}_f = 0.02 \times \dot{m}_a = 0.02 \times 3.873 = 0.07746 \ slug/sec$$

Total mass at the exit : $\dot{m}_f + \dot{m}_a = 3.95$ slug/sec

Momentum equation :

$$\Sigma F = (\dot{m}_f + \dot{m}_a) V_e - \dot{m}_a V_i = T$$ (pressure forces are neglected).

$$T = 3.95 \times 900 - 3.873 \times 500 = 1618.5 \ lb_f$$

$$\frac{T \times V_i}{550} = \text{Thrust horsepower}$$

$$\frac{1618.5 \times 500}{550} = 1471.4 \ hp.$$

4-11

A block is placed within a channel as shown. Calculate the force exerted by the flowing water on the block. The block and channel are both two meters wide normal to the paper.

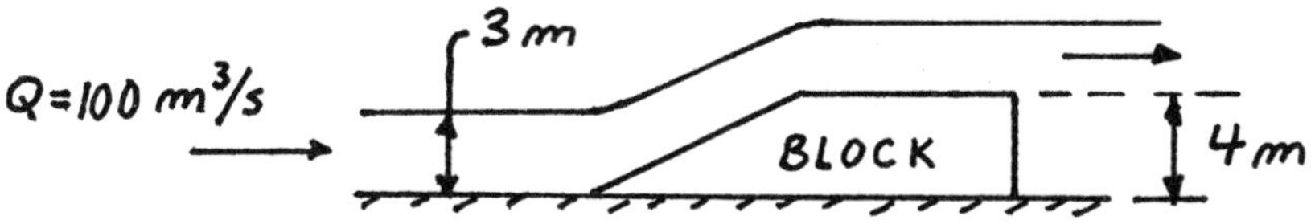

****

SOLVE FOR HEIGHTS AND VELOCITIES

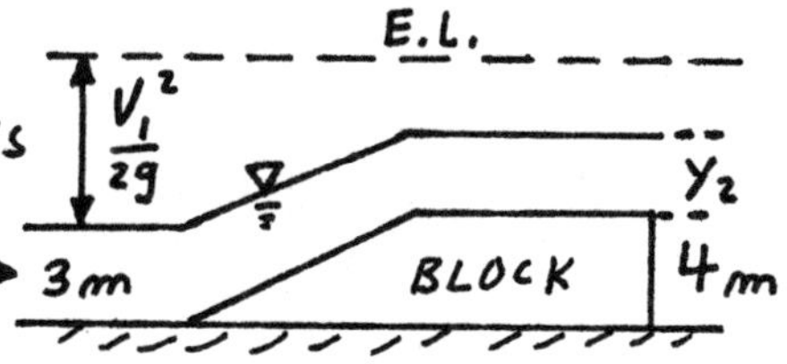

$$\frac{P}{\gamma} + \frac{V^2}{2g} + z = H$$

$$3 + \frac{V_1^2}{2g} = 4 + y_2 + \frac{V_2^2}{2g}$$

Subs. $\frac{Q}{A} = V \qquad A_1 = 6 \qquad A_2 = 2y_2$

$$3 + \frac{Q^2}{(6)^2}\frac{1}{2g} = 4 + y_2 + \frac{Q^2}{(2y_2)^2}\frac{1}{2g}$$

USE $Q = 100\ m^3/s$, $g = 9.81\ m/s^2$

SOLVE BY TRIAL AND ERROR $\qquad y_2 = \underline{\underline{3.66\ m}}$

SOLVE FOR UNKNOWN VALUES

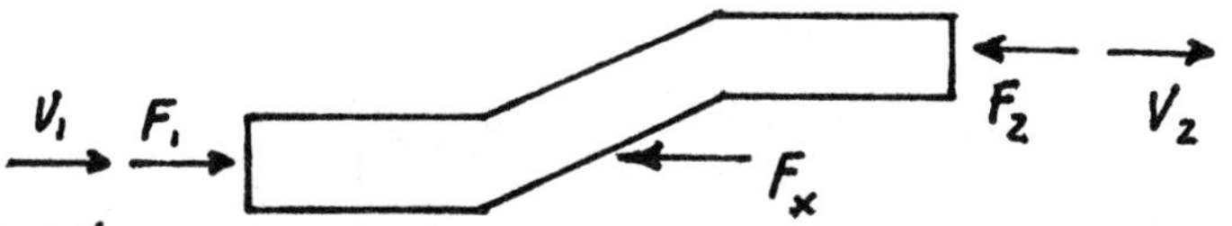

$$F_1 = \gamma h_c A = \gamma \frac{y_1}{2}(2y_1)$$

$$= (9.789\ kN/m^3)(3)^2 = \underline{\underline{88.1\ kN}} = \underline{\underline{88{,}100\ N}}$$

$$V_1 = \frac{Q_1}{A_1} = \frac{100\ m^3/s}{(3m)(2m)} = \underline{\underline{16.7\ m/s}} \qquad V_2 = \frac{Q_2}{A_2} = \frac{100\ m^3/s}{(2)(3.66)} = \underline{\underline{13.7\ m/s}}$$

$$F_2 = \gamma h_c A = \gamma \frac{y_2}{2}(2y_2) = (9.789\ kN/m^3)(3.66m) = \underline{\underline{35.8\ kN}} = \underline{\underline{35{,}800\ N}}$$

SOLVE FOR IMPULSE-MOMENTUM FORCE

$$\Sigma F_x = F_1 - F_x - F_2 = Q\rho(V_{OUT} - V_{IN})$$

$$\Sigma F_x = 88{,}100 - F_x - 35{,}800 = (100)(998.2)(13.7 - 16.7) \qquad F_x = \underline{\underline{351{,}760\ N}}$$

4-12

The sprinkler as shown will discharge water at 20 GPM. The rotating unit maintains a resisting torque of $0.01\omega^2$, (resisting motion). The nozzle inside diameter is 0.25 inch. Use moment of momentum principles to determine the RPM for the sprinkler.

****

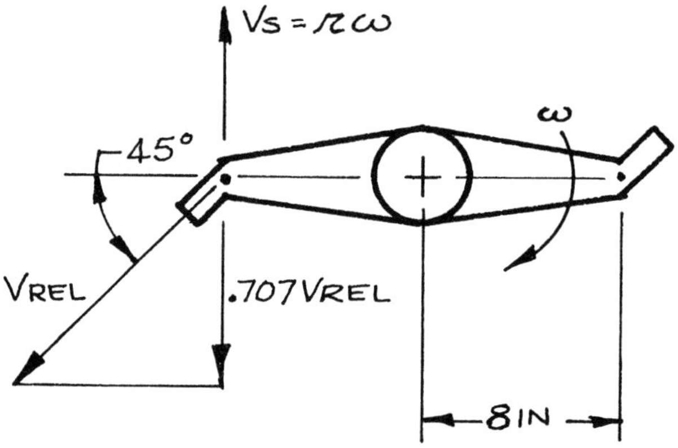

VECTOR TABLE

Vw =	Vs +	Vw/s
	rω	Q/A
	↓	↗45

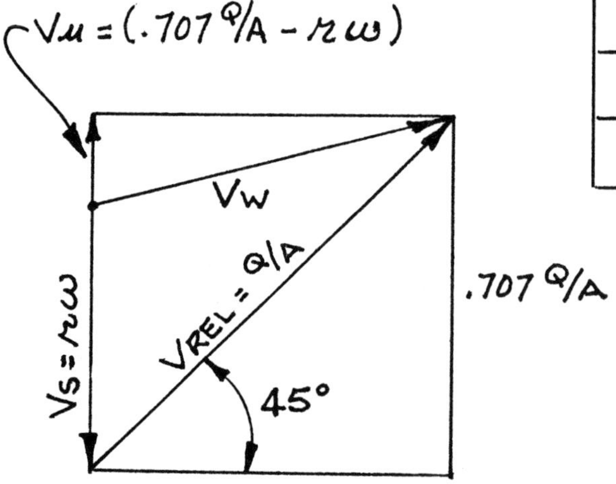

VECTOR DIAGRAM

TORQUE RESISTANCE $= 0.01\,\omega^2$

NOZZLE DIA (.25/12) FT

AREA $= \pi/4\,(D^2)$ FT^2

Q = 10 GPM/SIDE = 0.0222 CFS

Q/A = $V_{W/S}$ = FLOW RATE OF WATER (WRT) SPRINKLER

$V_S = r\omega$ = SPRINKLER VELOCITY

V_u AT LEVER ARM (8/12) PRODUCES TORQUE

Σ MOMENT OF MOMENTUMS = 0

$$\rho_1 Q_1 V_{u_1} R_1 + \rho_2 Q_2 V_{u_2} R_2 - .01\,\omega^2 = 0$$

FIND $V_{u_1} = .707\,V_{REL} - r\omega$

$$V_{u_1} = V_{u_2} = .707\,(Q/A) - r\omega$$

$$V_{u_1} = 46.212 - .666\,\omega \qquad (\gamma = \rho g)$$

$$2\,\frac{\gamma}{g}\,Q V_{u_1} R - .01\,\omega^2 = 0$$

$$2.6605 - .03823\,\omega - .01\,\omega^2 = 0$$

(OR) $\omega^2 + 3.823\,\omega - 266.05 = 0$

QUADRATIC EQUATION PROVIDES:

$14.5\ \frac{\text{RAD}}{\text{SEC}}$ = <u>138.6 RPM</u>

4-13

Water enters a horizontal y-tube joint through a 4 in. diameter pipe and exits through two 3 in. diameter pipe sections. If the entering velocity is 15 ft/s, find the exit velocities. For an entrance pressure of 60 psig and exit pressure of 40 psig, find the resultant force in the x-direction which is required to hold the joint in place.

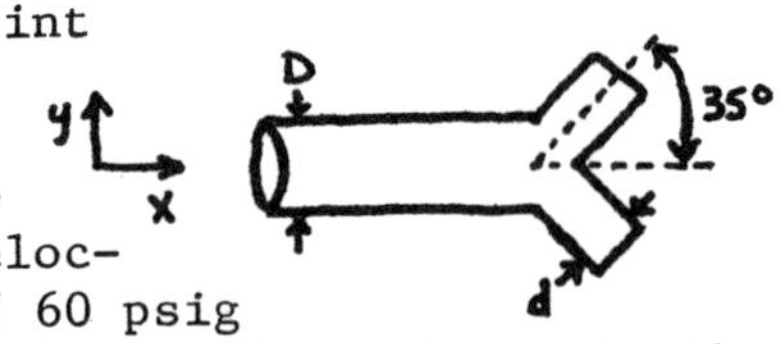

D= 4 in.
d= 3 in.

****

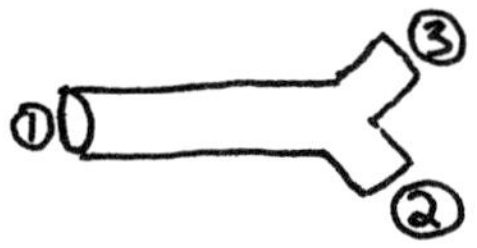

Assume:
1. Steady, incompressible flow.
2. Uniform properties over entrance and exits.
3. Both exit flows are identical.

Use conservation of mass to find exit velocities.

$$\frac{\partial}{\partial t}\iiint_{CV} \rho\, dV \;(\to 0) + \iint_{CS} \rho(\underset{\sim}{v}\cdot\underset{\sim}{n})dA = 0$$

$$\iint_{①} \rho(\underset{\sim}{v}\cdot\underset{\sim}{n})dA + \iint_{②} \rho(\underset{\sim}{v}\cdot\underset{\sim}{n})dA + \iint_{③} \rho(\underset{\sim}{v}\cdot\underset{\sim}{n})dA = 0$$

Normal velocity component is zero at the pipe wall.

$$-\rho_1|v_1|A_1 + 2\rho_2|v_2|A_2 = 0 \Rightarrow v_2 = \frac{v_1 A_1}{2A_2} = \frac{15\text{ ft/s}\,(4)^2}{2\,(3)^2} = 13.3\text{ ft/s}$$

($\rho_1 = \rho_2$ for incompressible flow)

Use the x-component of the momentum equation to find the resultant force R_x. Assume R_x is in the positive direction.

$$\Sigma F_x = \frac{\partial}{\partial t}\iiint_{CV} \rho v_x\, dV \;(\to 0) + \iint_{CS} \rho v_x(\underset{\sim}{v}\cdot\underset{\sim}{n})dA$$

$$R_x + P_1 A_1 - P_2 A_2 \cos 35° - P_3 A_3 \cos 35° = \iint_{①} \rho v_x(\underset{\sim}{v}\cdot\underset{\sim}{n})dA + \iint_{②} \rho v_x(\underset{\sim}{v}\cdot\underset{\sim}{n})dA + \iint_{③} \rho v_x(\underset{\sim}{v}\cdot\underset{\sim}{n})dA$$

Contributions at the exits ② and ③ may be combined. Solving for R_x:

$$R_x = 2P_2A_2\cos 35° - P_1A_1 - \rho_1 v_1|v_1|A_1 + 2\rho_2 v_2 \cos 35° |v_2| A_2$$

$$R_x = 2\left(40\,\frac{\text{lb}_f}{\text{in}^2}\right)\pi\,\frac{(3\text{in})^2}{4}\cos 35° - \left(60\,\frac{\text{lb}_f}{\text{in}^2}\right)\pi\,\frac{(4\text{in})^2}{4} - \left(1.94\,\frac{\text{slugs}}{\text{ft}^3}\right)(15\text{ ft/s})^2\,\pi\,\frac{(\frac{4}{12}\text{ft})^2}{4}$$
$$+\,2\left(1.94\,\frac{\text{slugs}}{\text{ft}^3}\right)(13.3\text{ ft/s})^2\cos 35°\,\pi\,\frac{(\frac{3}{12}\text{ft})^2}{4}$$

$$R_x = -301\text{ lb}_f$$

(The minus sign indicates the force is in the negative x-direction as one should expect.)

4-14

A horizontal jet of 50 mm diameter strikes a fixed vertical plate and is deflected through an angle of 90^0. Determine the force experienced by the plate if the jet velocity is 120 m/s. Assume density of water ρ = 1,000 kg/m^3.

****

Force experienced by plate, $F = \rho A V^2$

or, $F = 1000 \times \frac{\pi}{4}(0.050)^2 \times (120)^2 = 28{,}274$ N

or $F = 28.3$ kN

5
ENERGY PRINCIPLE

APPLICATIONS OF BERNOULLI'S AND ENERGY EQUATIONS

5-1

Water is flowing through a vertical pipe at a rate of 1 cfs in the upward direction. Determine the velocity and the pressure at the bottom end of the pipe if the pressure at the top end is atmospheric. The length of the pipe is 100 ft. Consider $\rho = 1.940$ slug/ft^3 and $g = 32.174$ ft/s^2. The head loss through the pipe is 10 ft of water. The pipe diameter is 4 in.

****

Considering the datum plane through the bottom of the pipe, the Bernoulli Equation for the real fluid flow is:

$$\frac{p_1}{\rho g} + \frac{V_1^2}{2g} + z_1 = \frac{p_2}{\rho g} + \frac{V_2^2}{2g} + z_2 + h_f$$

$$V_1 = V_2 = V = \frac{Q}{A} = \frac{1.00}{\frac{\pi}{4}\left(\frac{4}{12}\right)^2} = 11.459 \text{ ft/sec}$$

Then, $$\frac{p_1}{\rho g} = \frac{p_2}{\rho g} + z_2 + h_f - z_1 = 0 + 100 + 10 - 0$$

$$p_1 = 110\,\rho g = \frac{110 \times 1.940 \times 32.174}{144} = 47.680 \text{ psig}$$

5-2

Steam enters a turbine at 200 psia and 700°F with an inlet velocity of 25 ft/s. At the exit of the turbine, velocity is 35 ft/s and steam is at 300°F and 20 psia. Assuming adiabatic flow, determine the horsepower developed by this turbine if the flow rate of steam is 4 lb/s.

****

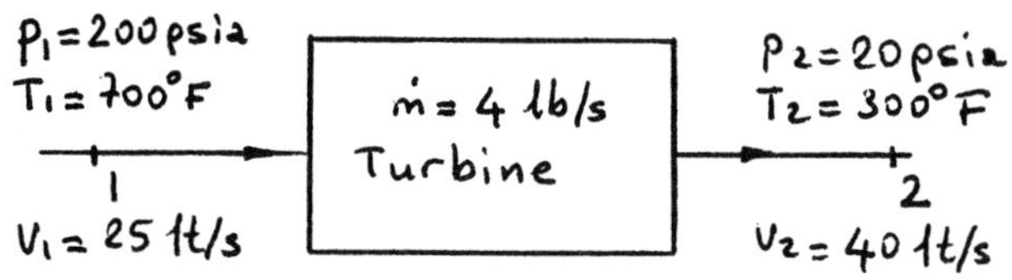

There is only 1 inlet and 1 outlet in this problem.

Therefore, $\dot{m}_1 = \dot{m}_2$ and flow is steady. Energy equation can be written as

$$-\dot{W}_s = \dot{m}_1 \left[(h_2 - h_1) + \left(\frac{V_2^2 - V_1^2}{2g_c}\right) + \frac{g(z_2 - z_1)}{g_c} \right]$$

Here, the last term can be neglected (system is horizontal and potential energy change is not significant). From steam tables (can be found in any thermodynamics text book) for states 1 and 2, we can read, $h_1 = 1374.3$ Btu/lbm , $h_2 = 1191.4$ Btu/lbm

Therefore $$-\dot{W}_s = 4\left[(1191.4 - 1374.3) + \frac{40^2 - 25^2}{2 \times 778 \times 32.2} + 0\right]$$

$$-\dot{W}_s = -731.6 + 0.08 = -731.52 \text{ Btu/s}$$

Therefore $$\dot{W}_s = 731.52 \text{ Btu/s}$$

or $$\dot{W}_s = 731.52 \times \frac{3600}{2545} = 1034.76 \text{ hp}$$

5-3

Ten cubic feet per second of water are flowing in the system shown. Determine the pressure gage reading and the velocity of the jet of water discharging at the nozzle. Draw the energy line and the hydraulic grade line.

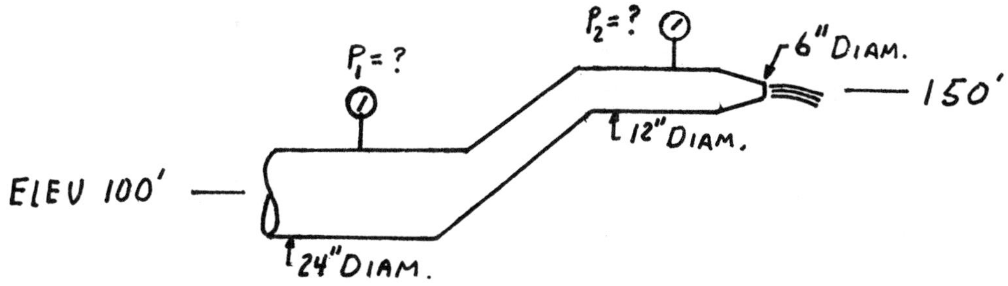

****

CALCULATE VELOCITY AT NOZZLE

$$V_3 = \frac{Q}{A_3} = \frac{10}{\left[\frac{\pi(6/12)^2}{4}\right]}$$

$V_3 = \underline{\underline{50.9}}$ fps

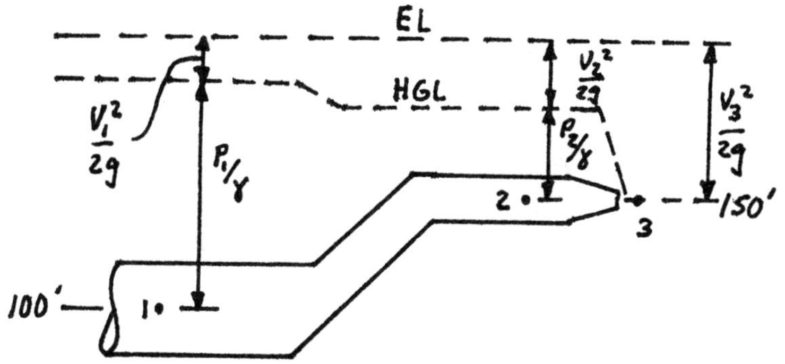

FIND ELEVATION OF ENERGY LINE

$$H = \frac{P}{\gamma} + \frac{V_3^2}{2g} + z_3 = 0 + \frac{(50.9)^2}{(2)(32.2)} + 150 = 0 + 40.2 + 150 = \underline{\underline{190.2 \text{ ft.}}}$$

CALCULATE PRESSURES AT POINT 1 AND POINT 3

$$V_1 = \frac{Q}{A_1} = \frac{10}{\pi(1)^2} = \underline{\underline{3.18 \text{ fps}}} \qquad V_2 = \frac{Q}{A_2} = \frac{10}{\pi(.5)^2} = \underline{\underline{12.7 \text{ fps}}}$$

$$190.2 = \frac{P_1}{62.4} + \frac{(3.18)^2}{(2)(32.2)} + 100 \qquad P_1 = 5619 \text{ psf} = \underline{\underline{39.0 \text{ psi}}}$$

$$190.2 = \frac{P_2}{62.4} + \frac{(12.7)^2}{(2)(32.2)} + 150 \qquad P_2 = 2352 \text{ psf} = \underline{\underline{16.3 \text{ psi}}}$$

5-4

If the pressure at Point A is 10 psi and the pressure at Point B is 25 psi when the volume flow rate of water is 1.57 cfs, find the direction of flow.

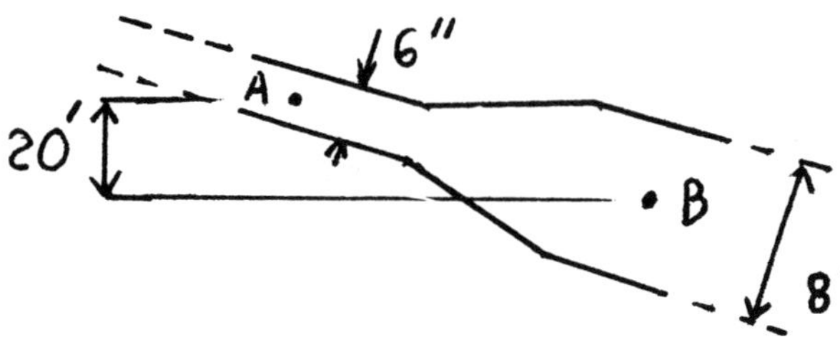

****

FROM CONTINUITY $\quad Q_A = Q_B = 1.57\ cfs$

$$V_A = \frac{Q_A}{A_A} = \frac{1.57\ FT^3/S}{\frac{\pi}{4}\left(\frac{6}{12}\right)^2 FT^2} = 8\ FT/S$$

$$V_B = \frac{Q_B}{A_B} = \frac{1.57\ FT^3/S}{\frac{\pi}{4}\left(\frac{8}{12}\right)^2 FT^2} = 4.5\ FT/S$$

ASSUME FLOW FROM A to B
WRITE the ENERGY EQUATION FROM A to B

$$\frac{P_A}{\gamma} + Z_A + \frac{V_A^2}{2g} = \frac{P_B}{\gamma} + Z_B + \frac{V_B^2}{2g} + h_f$$

$$\frac{10\frac{lb}{in^2}\left(144\frac{in^2}{FT^2}\right)}{62.4\ lb/FT^3} + 20' + \frac{8^2}{64.4} = \frac{25(144)}{62.4} + 0 + \frac{4.5^2}{64.4} + h_f$$

$$23.1' + 20' + 1' = 57.7' + 0 + 0.3' + h_f$$

$$h_f = 23.1 + 20 + 1 - 57.7 - 0.3 = -13.9'\ \text{(IMPOSSIBLE)}$$

Since $-h_f$ is not possible, flow is from B to A

5-5

A liquid (sg=.86) with a vapor pressure of 3.8 psia flows through a horizontal constriction as shown in the figure. Atmospheric pressure is 26.7 " Hg. Find the flow rate at which cavitation would occure.

Ignore any head loss through the constriction.

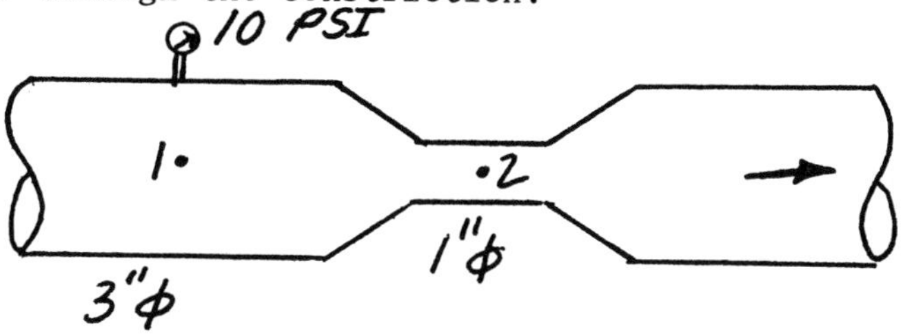

****

$\gamma = .86\,(62.4) = 53.66 \text{ lb/ft}^3$

$Q = A_1 V_1 = A_2 V_2 \quad ; \quad V_2 = 9 V_1$

Atmospheric Pressure $= \dfrac{26.7}{29.9}(14.7) = 13.1$ psia

Bernoulli's Eqn ① to ②; use absolute pressure:

$$\frac{P_1}{\gamma} + \cancel{z_1} + \frac{V_1^2}{2g} = \frac{P_2}{\gamma} + \cancel{z_2} + \frac{V_2^2}{2g}$$

$$\frac{(13.1 + 10)\,144}{53.66} + \frac{V_1^2}{2(32.2)} = \frac{3.8(144)}{53.66} + \frac{(9V_1)^2}{2(32.2)}$$

$$V_1 = 6.46 \text{ ft/sec}$$

$Q = A_1 V_1 = \dfrac{\pi}{4}\left(\dfrac{3}{12}\right)^2 (6.46) =$ | $.317 \dfrac{ft^3}{s}$ | ANS |

ENERGY GRADE LINE

5-6

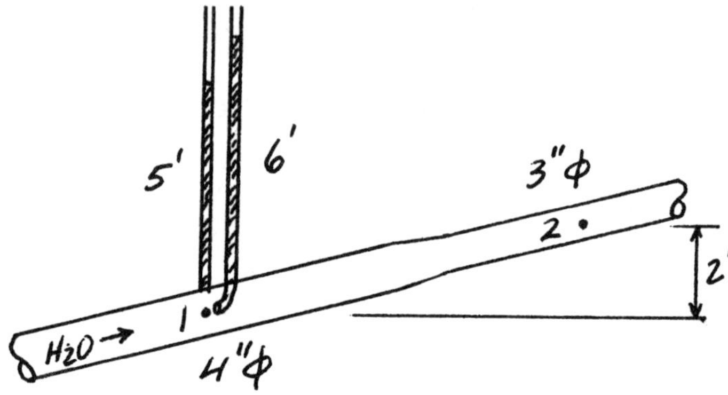

Water flowing through the pipe gives a reading of 5' on the piezometric tube and 6' on the pitot tube as shown in the figure. Find V_1, Q, $\frac{P_2}{\gamma}$ and $\frac{V_2^2}{2g}$. Assume no friction head loss.

Sketch the total head line (energy line) and the hydraulic grade line from point 1 to point 2.

****

$$\frac{V_1^2}{2g} = 1' \; ; \quad V_1 = \sqrt{2(32.2)(1)} = \boxed{8.02 \text{ ft/s} \quad \text{ANS}}$$

$$V_2 = \left(\frac{4}{3}\right)^2 (8.02) = 14.27 \text{ ft/s} \; ; \quad \frac{V_2^2}{2g} = \frac{(14.27)^2}{2(32.2)} = \boxed{3.16' \quad \text{ANS}}$$

$$Q = A_1 V_1 = \frac{\pi}{4}\left(\frac{4}{12}\right)^2 (8.02) = (.0873 \text{ ft}^2) 8.02 = \boxed{.700 \text{ ft}^3/\text{s} \quad \text{ANS}}$$

$$\frac{P_1}{\gamma} + z_1 + \frac{V_1^2}{2g} = \frac{P_2}{\gamma} + z_2 + \frac{V_2^2}{2g} \qquad (z_1 = 0)$$

$$6 = \frac{P_2}{\gamma} + 2 + 3.16 \; ; \quad \frac{P_2}{\gamma} = 6 - 2 - 3.16 = \boxed{.84' \quad \text{ANS.}}$$

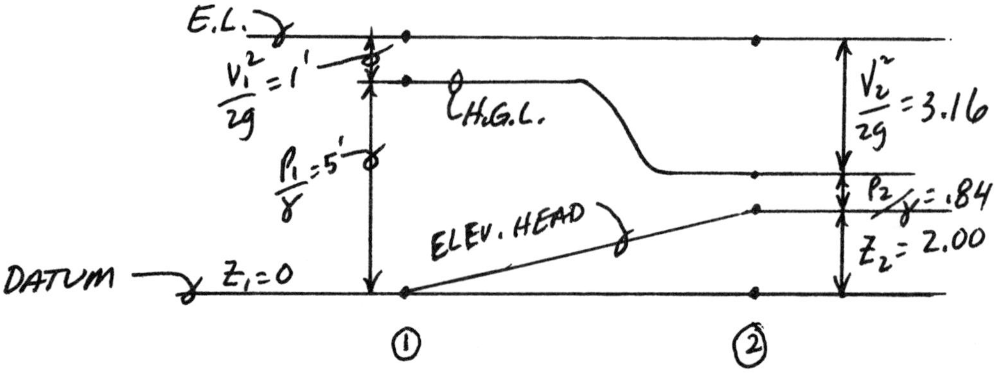

5-7

Water is flowing through a horizontal pipe of uniform diameter. Determine the energy gradient if the length of the pipe is 1,000 ft. Assume the head loss is 20 ft of water.

****

The Bernoulli Equation, as applied to real fluid flow is:

$$\frac{p_1}{\rho g} + \frac{V_1^2}{2g} + z_1 = \frac{p_2}{\rho g} + \frac{V_2^2}{2g} + z_2 + h_f$$

Then,

$$h_f = \left(\frac{p_1}{\rho g} + \frac{V_1^2}{2g} + z_1\right) - \left(\frac{p_2}{\rho g} + \frac{V_2^2}{2g} + z_2\right)$$

$$\text{The Energy Gradient} = \frac{dh_f}{dL} = \frac{20}{1000} = 0.02$$

PUMPS AND TURBINES IN THE SYSTEM

5-8

A turbine extracts power from water in a reservoir. The water drops 150 ft. from the reservoir surface in a 84 inch diameter penstock to the discharge. With a velocity in the penstock of 15 feet per second and turbine efficiency of 70 percent, compute the amount of energy that will be delivered to the power lines in kilowatts.

****

CALCULATE FLOW RATE AT DISCHARGE. DETERMINE ENERGY EXTRACTED BY TURBINE.

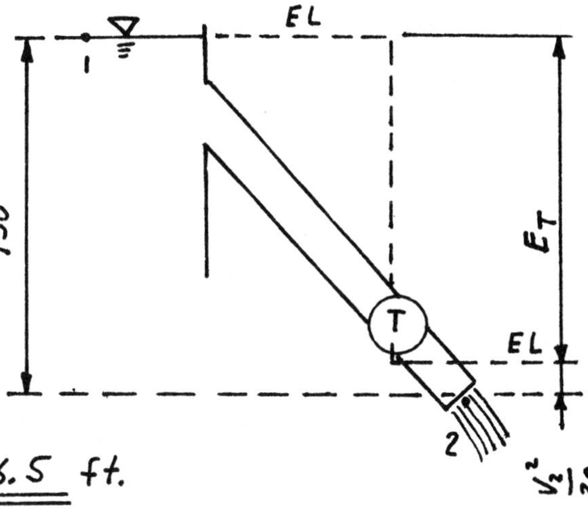

$$Q = AV = \frac{\pi (7)^2}{4}(15) = \underline{\underline{577 \text{ cfs}}}$$

$$E_T = 150 - \frac{V_2^2}{2g} = 150 - \frac{(15)^2}{(2)(32.2)}$$

$$E_T = 150 - 3.49 = 146.51 = \underline{\underline{146.5}} \text{ ft.}$$

CALCULATE USEFUL POWER EXTRACTED BY TURBINE

$$P = \frac{Q\gamma E_T \eta_T}{550} = \frac{(577 cfs)(62.4\ lb/ft^3)(146.5\ ft)(0.70)}{550\ \frac{ft\text{-}lb}{s}/hp} = 6713\ hp$$

$$P = 6713\ hp \times 0.746\ kW/hp = 5008\ KW$$

5-9

Determine the power and head developed by a pump that will be used to empty the tank as shown while discharging to an elevated exit port. The pump must empty the tank in 1 minute. (Assume the static inlet head and flow rate are constant.)

****

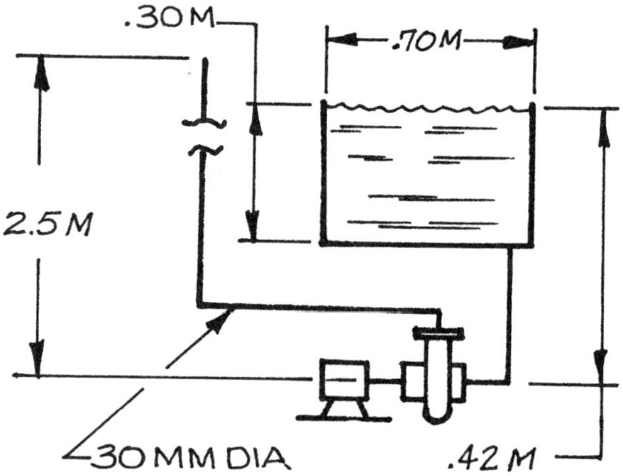

VOLUME OF FLUID:

$$\frac{\pi}{4}(.70)^2(.3) = 0.115\ M^3$$

FLOW RATE: (Q/t)

$$\frac{0.115}{60} = 0.001924\ M^3/s$$

DETERMINE HEAD REQUIRED BY THE PUMP (METERS)

$$\frac{P_1}{\gamma} + \frac{V_1^2}{2g} + Z_1 + H_P = \frac{P_2}{\gamma} + \frac{V_2^2}{2g} + Z_2$$

$$0 + 0 + .42 + H_P = 0 + \frac{\left[\frac{.001924}{\pi(.03)^2/4}\right]^2}{2g} + 2.5$$

$$\underline{H_P = 2.46\ M}$$

$$POWER = \gamma Q H_P = (9.81)(.001924)(2.46) = \underline{46.3\ W}$$

5-10

Water is flowing through the horizontal pipe system shown below. Determine the pump horsepower required to maintain a flow of 30 cfs. Disregard friction losses.

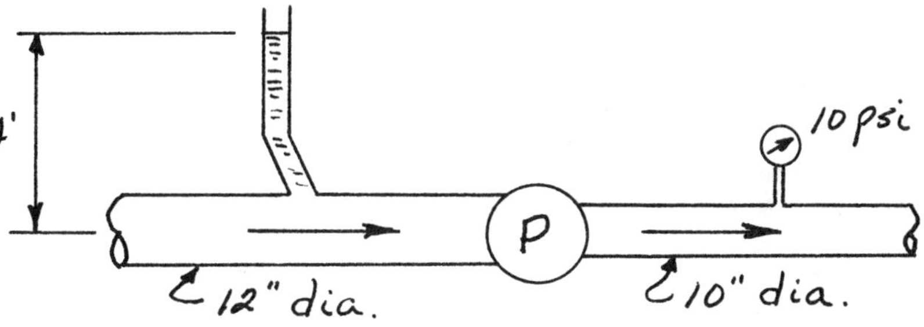

****

Let point ① be on intake side of pump.
Let point ② be on discharge side of pump.

$$\frac{V_1^2}{2g} + z_1 + \frac{P_1}{\gamma} + E_M = \frac{V_2^2}{2g} + z_2 + \frac{P_2}{\gamma}$$

$$V_1 = \frac{Q}{A_1} = \frac{30\ \text{ft}^3/\text{sec}}{\frac{\pi}{4}(1\ \text{ft})^2} = 38.2\ \text{ft/sec}$$

$$V_2 = \frac{Q}{A_2} = \frac{30\ \text{ft}^3/\text{sec}}{\frac{\pi}{4}\left(\frac{10}{12}\ \text{ft}\right)^2} = 55.0\ \text{ft/sec}$$

$z_1 = z_2$ since pipe system is horizontal

$$E_m = \frac{V_2^2 - V_1^2}{2g} + \frac{P_2}{\gamma} - \frac{P_1}{\gamma}$$

$$E_m = \frac{(55.0\ \text{ft/sec})^2 - (38.2\ \text{ft/sec})^2}{2 \times 32.2\ \text{ft/sec}^2} + \frac{10\ \frac{\text{lb}}{\text{in}^2}}{62.4\ \frac{\text{lb}}{\text{ft}^3}} \times \frac{144\ \text{in}^2}{\text{ft}^2} - 4\ \text{ft}$$

$$E_m = (24.3 + 23.1 - 4)\ \text{ft} = 43.4\ \text{ft}$$

$$\text{Power} = \frac{Q\gamma E_m}{550} = \frac{\left(30\ \frac{\text{ft}^3}{\text{sec}}\right)\left(62.4\ \frac{\text{lb}}{\text{ft}^3}\right)(43.4\ \text{ft})}{550\ \frac{\text{ft-lb}/\text{sec}}{\text{Hp}}} = \underline{\underline{147.7\ \text{Hp}}}$$

5-11

A fluid with a unit weight of 78.5 pcf is flowing at a rate of 14.5 cfs. What HP is the pump? Ignore any losses.

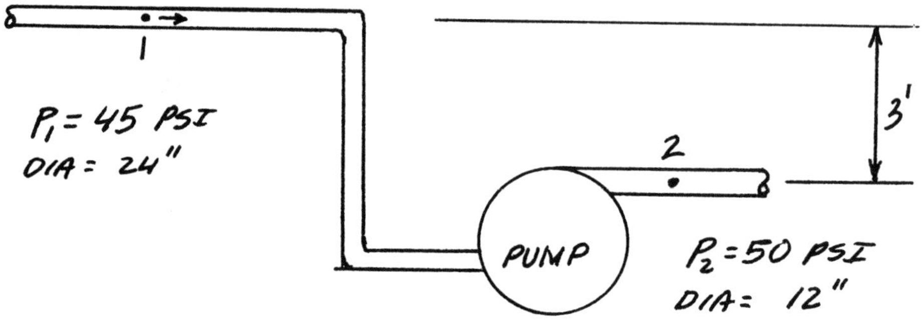

****

$$\frac{P_1}{\gamma} + Z_1 + \frac{V_1^2}{2g} + h_a = \frac{P_2}{\gamma} + Z_2 + \frac{V_2^2}{2g}$$

$$V_1 = \frac{Q}{A} = \frac{14.5}{\frac{\pi(2^2)}{4}} = 4.62 \text{ ft/s} ; \quad V_2 = \frac{14.5}{\frac{\pi(1^2)}{4}} = 18.46 \text{ ft/s}$$

$$\frac{45 \times 144}{78.5} + 3 + \frac{(4.62)^2}{2 \times 32.2} + h_a = \frac{50 \times 144}{78.5} + 0 + \frac{(18.46)^2}{2 \times 32.2}$$

$$h_a = 11.1 \text{ ft.}$$

$$P = \gamma h_a Q = \frac{(78.5)(11.1)(14.5)}{550} = \boxed{23 \text{ HP}}$$

5-12

Water flows through the turbine at a volume flow rate of $Q = 10\ ft^3/sec$. The pressure at point A is $P_A = 22$ psig. The pressure at point B is $P_B = -5.0$ psig. The power output from the turbine is $P_o = 60$ HP. Compute the efficiency, e_m, of the turbine. There are no friction losses.

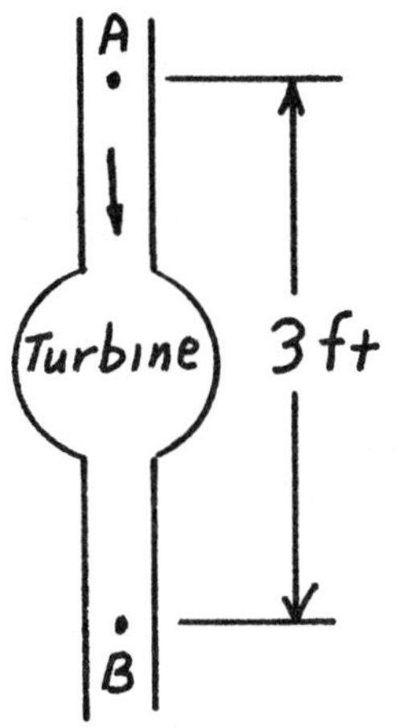

$P_A = 22\ psig$

$P_B = -5.0\ psig$

****

Apply energy equation

$$\frac{P_A}{\gamma} + Z_A + \frac{V_A^2}{2g} - h_R = \frac{P_B}{\gamma} + Z_B + \frac{V_B^2}{2g}$$

$$\frac{\left(22\frac{lb}{in^2}\right)\left(\frac{144\ in^2}{ft^2}\right)}{62.4\ lb/ft^3} + 3\ ft - h_R = \frac{\left(-5\frac{lb}{in^2}\right)\left(144\frac{in^2}{ft^2}\right)}{62.4\ lb/ft^3}$$

$$h_R = 50.8\ ft + 11.5\ ft + 3\ ft$$

$$h_R = 65.3\ ft \qquad P_R = h_R \gamma Q = (65.3\ ft)\left(62.4\frac{lb}{ft^3}\right)\left(10\frac{ft^3}{s}\right)$$

$$P_R = \left(40747\ \frac{ft\text{-}lb}{s}\right)\left(\frac{1\ HP}{550\ ft\text{-}lb/s}\right)$$

$$P_R = 74\ HP$$

$$e_m = P_o/P_R = 60\ HP/74\ HP$$

$$\underline{\underline{e_m = 0.81 = 81\%}}$$

5-13

A hydraulic turbine operates between a headwater and tailwater pool with a 170 foot difference in elevation. Determine the flow rate (in cfs) through a 90% efficient turbine with a power output of 800 horsepower and a system head loss of 20 feet.

****

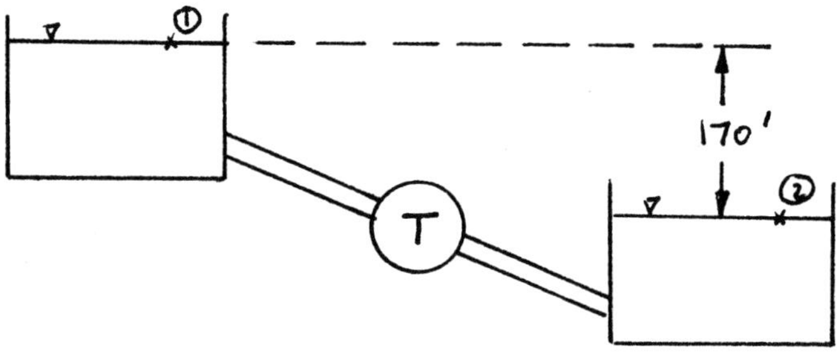

Bernoulli balance between ① & ②:

$$\frac{V_1^2}{2g} + z_1 + \frac{p_1}{\gamma} = \frac{V_2^2}{2g} + z_2 + \frac{p_2}{\gamma} + h_L + H_T$$

$$H_T = (z_1 - z_2) - h_L$$

$$= 170' - 20' = 150'$$

System horsepower, $hp = \frac{800}{0.9} = 889 = \frac{Q\gamma H_T}{550}$

$$Q = \frac{889\,(550)\ \text{ft}\cdot\text{lb/s}}{62.4\ \frac{\text{lb}}{\text{ft}^3}\,(150\ \text{ft})} = \underline{\underline{52.2\ \text{cfs}}}$$

5-14

A new design of a fuel pump requires it to pump (1) liter of fuel in 40 seconds with a suction pressure of 150 mm of mercury vacuum and a discharge pressure of 30 KPa. If the pump is 50 percent efficient, calculate power drawn from the engine. The specific weight of fuel is 72% of the weight of water. (metric)

****

CONVERT LITERS TO M^3 $\quad 1\ell = 0.001\ M^3$

SPECIFIC GRAVITY OF FUEL = 0.72

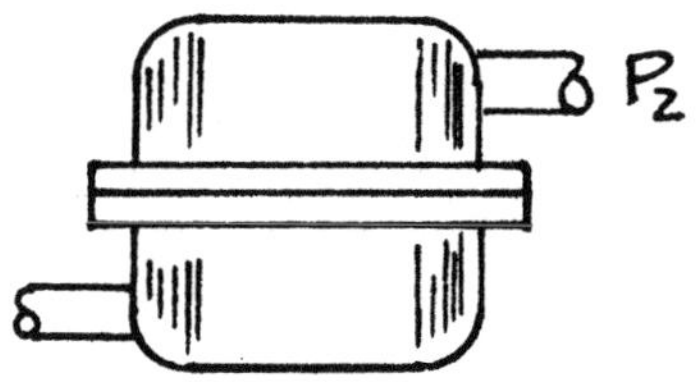

USE ENERGY EQUATION TO FIND HEAD REQUIRED BY PUMP

$$\frac{P_1}{\gamma} + \frac{V_1^2}{2g} + Z_1 + H_P = \frac{P_2}{\gamma} + \frac{V_2^2}{2g} + Z_2$$

$$V_1 = V_2 \, , \quad Z_1 = Z_2$$

$$H_P = \frac{P_2 - P_1}{\gamma} = \frac{30 - [-.15(13.6)(9.81)]}{0.72(9.81)}$$

$$H_P = 7.09\ M$$

$$\text{POWER} = \gamma Q H_P = 0.72(9.81)\frac{10^{-3}}{40}(7.09)$$

POWER TO FLUID = <u>1.25 W</u>

$$\text{EFFICIENCY } \eta = \frac{\text{POWER OUT}}{\text{POWER IN}} = 0.50$$

<u>POWER IN TO PUMP = 2.50 W</u>

5-15

Water is being supplied to a turbine at a rate of 50 cfs and with an inlet pressure of 500 ft of water. Determine the power output of the turbine if it discharges the water to the open atmosphere. Assume the inlet and outlet pipe diameters of the turbine are the same. Consider the turbine to be 87% efficient. The density of water is 1.940 slug/ft^3.

× 32.2 → 62.4 lb/ft^3

****

Bernoulli Equation for flow through a turbine is:

$$\frac{p_1}{\rho g} + \frac{V_1^2}{2g} + z_1 = \frac{p_2}{\rho g} + \frac{V_2^2}{2g} + z_2 + h_t$$

$$500 + \frac{V^2}{2g} + 0 = 0 + \frac{V^2}{2g} + 0 + h_t$$

or, head on turbine, $h_t = 500$ ft of water

Power of Turbine $= P = \dfrac{\rho g Q h_t}{550}$

or, $P = \dfrac{1.940 \times 32.174 \times 50 \times 500}{550}$

or, $P = 2{,}837$ h.p.

Power Output of Turbine = P × efficiency

$= 2837 \times 0.87 = 2{,}468$ h.p.

5-16

The suction and delivery gages of a pump read pressures of -5 m and 40 m of water, respectively. The suction and delivery pipes have diameters of 150 and 125 mm, respectively. Determine the power delivered by the pump. Also, determine its power input if its efficiency is 82%. The flow rate is 0.1 m^3/s and the total head loss involved in the pumping system is 10 m of water. Consider ρ = 1 000 kg/m^3 and g = 9.810 m/s^2. Assume the pump centerline and the centers of both gages to be located at the same datum plane.

****

Considering the datum plane through the pump centerline, the Bernoulli Equation for the pump is:

$$\frac{p_1}{\rho g} + \frac{V_1^2}{2g} + z_1 = \frac{p_2}{\rho g} + \frac{V_2^2}{2g} + z_2 + h_f - h_p$$

$$(-5) + \frac{\left(\frac{0.1}{\pi (0.150)^2/4}\right)^2}{2g} + 0 = 40 + \frac{\left(\frac{0.1}{\pi (0.125)^2/4}\right)^2}{2g} + 0 + 10 - h_p$$

or, $-5 + 111.026 + 0 = 40 + 159.878 + 0 + 10 - h_p$

or, head developed by pump $= h_p = 103.852$ m

Power delivered by Pump $= P = \dfrac{\rho g Q h_p}{1000}$ kW

or, $P = \dfrac{1000 \times 9.810 \times 0.1 \times 103.852}{1\,000} = 102$ kW

Power Input $= \dfrac{P}{\text{efficiency}} = \dfrac{102}{0.82} = 124.$ kW

5-17

A pump is to deliver water to a height of 30 ft. at a maximum flowrate of 600 gal/min and a pressure of 80 psig through a 4 in. diameter pipe. If the pump intake is 6 in. diameter pipe and water is available at 40 psig, find the horsepower required. Assume the pump is 60% efficient; neglect heat losses.

****

Energy equation: $$\frac{\delta Q}{dt}^{\nearrow 0(2)} - \frac{\delta W_\mu}{dt}^{\nearrow 0(4)} = \iint_{CS} \left(e+\frac{P}{\rho}\right)\rho(\underset{\sim}{v}\cdot\underset{\sim}{n})\,dA + \frac{\partial}{\partial t}\iiint_{CV} \rho e\,dV^{\nearrow 0(1)} + \frac{\delta W_s}{dt}$$

(1) Assume steady, incompressible flow.
(2) Neglect heat losses.
(3) No change in internal energy
(4) No viscous work across the control surface

$$\frac{\delta W_s}{dt} = -\iint_{CS}\left(e+\frac{P}{\rho}\right)\rho(\underset{\sim}{v}\cdot\underset{\sim}{n})\,dA = -\iint_{①}\left(e+\frac{P}{\rho}\right)\rho(\underset{\sim}{v}\cdot\underset{\sim}{n})\,dA - \iint_{②}\left(e+\frac{P}{\rho}\right)\rho(\underset{\sim}{v}\cdot\underset{\sim}{n})\,dA$$

① entrance
② exit

with specific energy $e = u + \frac{v^2}{2} + gz$

$$\frac{\delta W_s}{dt} = \left(\frac{v_1^2}{2} + gz_1 + u_1 + \frac{P_1}{\rho_1}\right)\rho_1 v_1 A_1 - \left(\frac{v_2^2}{2} + gz_2 + u_2 + \frac{P_2}{\rho_2}\right)\rho_2 v_2 A_2$$

$\rho_1 = \rho_2$ by (1) above; continuity equation implies: $\rho v_1 A_1 = \rho v_2 A_2 = \rho Q$

$$\frac{\delta W_s}{dt} = \left[\frac{v_1^2}{2} - \frac{v_2^2}{2} + g(z_1 - z_2) + (u_1 - u_2)^{\nearrow 0(3)} + \frac{P_1 - P_2}{\rho}\right]\rho Q$$

The velocity at the entrance can be found from the volumetric flowrate and area perpendicular to the flow.

$$v_1 = \frac{Q}{A_1} = \frac{4Q}{\pi D_1^2} = \frac{4(600\text{ gal/min})\left(\frac{1\text{ min}}{60\text{ s}}\right)}{\pi\,(.5\text{ ft})^2\,(7.49\text{ gal/ft}^3)} = 6.8\text{ ft/s}\ ;\ v_2 = \frac{Q}{A_2} = 15.3\text{ ft/s}$$

Substituting,

$$\frac{\delta W_s}{dt} = \left[\frac{(6.8\text{ ft/s})^2}{2} - \frac{(15.3\text{ ft/s})^2}{2} + (32.2\text{ ft/s}^2)(-30\text{ ft.}) + \frac{\left(40\frac{\text{lb}_f}{\text{in}^2} - 80\frac{\text{lb}_f}{\text{in}^2}\right)\left(\frac{144\text{ in}^2}{\text{ft}^2}\right)}{1.94\text{ slugs/ft}^3}\right]$$

$$\times\frac{(1.94\text{ slugs/ft}^3)(600\text{ gal/min})}{(7.49\text{ gal/ft}^3)(60\text{ s/min})} = -10{,}400\text{ ft-lb}_f/\text{s}$$

$$\frac{\delta W_s}{dt} = -10{,}400\,\frac{\text{ft-lb}_f}{\text{s}}\cdot\frac{1\text{ Hp}}{550\text{ ft-lb}_f/\text{s}} = -19\text{ Hp}\ ;\ 60\%\text{ efficiency} \Rightarrow 32\text{ Hp.}$$

6
DIMENSIONAL ANALYSIS

BUCKINGHAM PI THEOREM

6-1

In the March 1982 issue of the AIChE Journal, Vasalos and coworkers reported experimental findings for particle holdup times in a synthetic fuels reactor. They showed that the Reynolds number at the terminal velocity of a spherical particle was related to the Galileo number. Use the Buckingham Pi Theorem to find these dimensionless groups if the 5 important parameters are diameter D, density ρ, viscosity μ, velocity v and the buoyant force per unit volume $(\rho_s - \rho)g$. Choose your repeating or core parameters in the order given[that is, begin with diameter, D, then density, ρ, and so on until the appropriate number is obtained].

****

Parameters: $D \quad \rho \quad \mu \quad v \quad (\rho_s - \rho)g \qquad n = 5$ parameters

Use m, L, and t as primary dimensions (r = 3). In terms of primary dimensions, units for these parameters are:

$$L \quad \frac{m}{L^3} \quad \frac{m}{Lt} \quad \frac{L}{t} \quad \frac{m}{L^2t^2}$$

Three repeating parameters should be selected: D, ρ, and μ
This means 5-3=2 dimensionless π-groups, which are obtained by combining the set of repeating parameters with each of the other remaining parameters one at a time.

$$\pi_1 = D^a \rho^b \mu^c v \qquad\qquad \pi_2 = D^e \rho^f \mu^h (\rho_s - \rho) g$$

For these to be dimensionless, we must have:

$$(L)^a \left(\frac{m}{L^3}\right)^b \left(\frac{m}{Lt}\right)^c \left(\frac{L}{t}\right) = m^0 L^0 t^0 \qquad (L)^e \left(\frac{m}{L^3}\right)^f \left(\frac{m}{Lt}\right)^h \left(\frac{m}{L^2 t^2}\right) = m^0 L^0 t^0$$

Exponent equation for

$$\left.\begin{aligned} m&: \; b+c=0 \\ L&: \; a-3b-c+1=0 \\ t&: \; -c-1=0 \end{aligned}\right\} \Rightarrow \begin{aligned} b&=1 \\ a&=1 \\ c&=-1 \end{aligned} \qquad \left.\begin{aligned} m&: \; f+h+1=0 \\ L&: \; e-3f-h-2=0 \\ t&: \; -h-2=0 \end{aligned}\right\} \Rightarrow \begin{aligned} f&=1 \\ e&=3 \\ h&=-2 \end{aligned}$$

Hence,

$$\pi_1 = D^a \rho^b \mu^c v = \frac{\rho v D}{\mu} \; \binom{\text{Reynold's}}{\text{Number}} \qquad \pi_2 = D^e \rho^f \mu^h (\rho_s - \rho) g = \frac{\rho D^3 (\rho_s - \rho) g}{\mu^2} \; \binom{\text{Galileo}}{\text{Number}}$$

DIMENSIONLESS NUMBERS

6-2

The Department of Defense wants to model an intercontinental ballistic missile (ICBM) in a wind tunnel. Derive a basic functional relationship for the strike force (F) of such a missile if the force is considered a function of missile velocity (V) and diameter (D) as well as air density (ρ), viscosity (μ) and bulk modulus (E_V). Use V, D and ρ as repeating variables. What common dimensionless parameters must be equivalent in the model and prototype to insure dynamic similitude?

****

$$F = f(\underline{V}, \underline{D}, \underline{\rho}, \mu, E_v)$$

Use MLT system and Hunsaker-Rightmire method

$$\left.\begin{aligned} V &= \frac{L}{T} \\ D &= L \\ \rho &= \frac{M}{L^3} \end{aligned}\right\} \quad L = D; \quad T = \frac{D}{V}; \quad M = \rho D^3$$

$$F = \frac{ML}{T^2} = \frac{(\rho D^3)D}{D^2/V^2} = \rho D^2 V^2 \quad \Rightarrow \quad \pi_1 = \frac{F}{\rho D^2 V^2}$$

$$\mu = \frac{M}{LT} = \frac{\rho D^3}{D(D/V)} = \rho D V \quad \Rightarrow \quad \pi_2 = \frac{\mu}{\rho D V}$$

$$\pi_2^{-1} = \frac{\rho D V}{\mu} = \underline{\mathbb{R}_e}$$

$$E_V = \frac{M}{LT^2} = \frac{\rho D^3}{D(D^2/V^2)} = \rho V^2 \quad \Rightarrow \quad \pi_3 = E_V/\rho V^2$$

$$\pi_3^{-1/2} = \frac{V}{\sqrt{E_V/\rho}} = \underline{M_N}$$

$$\underline{F = \rho D^2 V^2 f(\mathbb{R}_e, M_N)}$$

SIMILITUDE AND MODEL ANALYSIS

6-3

Water at 60°F flows in a pipe of 3 inches diameter with a velocity of 15 ft/s. If air flows in another pipe at the same temperature with a velocity of 20 ft/s, determine the required pipe diameter in order that these two flows become dynamically similar.What is the required diameter for the second pipe if the air flow is at 150°F ?

****

For dynamic similarity, Reynolds numbers should be equal :

$$Re_1 = \frac{V_1 \, d_1}{\nu_1} = Re_2 = \frac{V_2 \, d_2}{\nu_2} \qquad 1: \text{water}, \; 2: \text{air}$$

For 60°F: $$\frac{15 \times 3/12}{1.22 \times 10^{-5}} = \frac{20 \times d_2/12}{1.58 \times 10^{-4}} \quad \rightarrow d_2 = 29.14 \text{ inches}.$$

For 150°F: $$\frac{15 \times 3/12}{1.22 \times 10^{-5}} = \frac{20 \times d_2/12}{2.07 \times 10^{-4}} \quad \rightarrow d_2 = 38.17 \text{ inches}.$$

6-4

A 1:50 scale model of the Betzville (PA) Dam is desired by the U.S. Army Corps of Engineers to help analyze problems that have developed in the prototype. The dam has the following characteristics:

Dam width:	4000'
Spillway length:	2500'
Spillway design flow:	1000 cfs

Determine the minimum room dimensions required to house the model and the model pumping capacity required.

****

Length ratio $(l_r) = 50$

$$\therefore \text{Room size} = \frac{4000'}{50} \times \frac{2500'}{50} = \underline{\underline{80' \times 50'}}$$

Froude No. problem: $\mathbb{F}_r = \dfrac{V}{\sqrt{gl}}$

$$\mathbb{F}_{r_p} = \mathbb{F}_{r_m}$$

$$\frac{V_p}{\sqrt{g_p l_p}} = \frac{V_m}{\sqrt{g_m l_m}}$$

Velocity ratio, $V_r = \dfrac{V_p}{V_m} = \dfrac{\sqrt{l_p}}{\sqrt{l_m}}$

Flow ratio, $Q_r = V_r A_r$

$$Q_r = \frac{V_p}{V_m}\frac{A_p}{A_m} = \frac{\sqrt{l_p}}{\sqrt{l_m}}\frac{l_p^2}{l_m^2} = \left(\frac{l_p}{l_m}\right)^{5/2}$$

$$\frac{Q_p}{Q_m} = l_r^{5/2} = 50^{5/2} = 17680$$

$$Q_m = \frac{1000 \text{ cfs}}{17680} = \underline{\underline{.0566 \text{ cfs}}}$$

6-5

A prototype spillway is 50 m high and has a crest which is 100 m long. The head over the crest is 2 m. For a discharge of 2 800 m^3/s of water over the prototype spillway, determine the discharge over the model spillway. The scale ratio of prototype to model is 36:1.

****

The discharges of prototype and model spillways are related as:

$$\frac{Q_m}{Q_p} = \frac{(L H^{3/2})_m}{(L H^{3/2})_p}$$

$$Q_m = \frac{L_m}{L_p} \cdot \frac{H_m^{3/2}}{H_p^{3/2}} \cdot Q_p = \frac{1}{36} \times \left(\frac{1}{36}\right)^{3/2} (2800)$$

$$Q_m = 0.360 \ m^3/s$$

6-6

A race car is planned to travel at 120 km/hr at standard atmospheric conditions (25°C and 1 atm.).Its 1:12 scale model is to be tested in a wind tunnel operating with air at 25°C.In order that the model and the car have the same Mach number and Reynolds number,what must be the tunnel pressure?Calculate the tunnel pressure for a tunnel temperature of 50°F and compare the results.

****

In the first case, for the same temperature $T_1 = T_2$

$$c_1 = \sqrt{kRT_1} = c_2 = \sqrt{kRT_2} \qquad \text{speed of sound is same.}$$

Therefore, for the same Mach number $M_1 = \frac{V_1}{c_1} = M_2 = \frac{V_2}{c_2}$,

we must have $V_1 = V_2$, i.e. velocities should be same.

In order to have dynamic similarity, Reynolds numbers should be similar:

$$Re_1 = \frac{V_1 L_1 \rho_1}{\mu_1} = Re_2 = \frac{V_2 L_2 \rho_2}{\mu_2} \qquad (\text{also, } \mu_1 = \mu_2 \text{, since } T_1 = T_2)$$

Therefore: $L_1 \rho_1 = L_2 \rho_2 \rightarrow \frac{L_2}{L_1} = \frac{\rho_1}{\rho_2} = \frac{1}{12}$

Perfect gas law: $p_1 = \rho_1 R_1 T_1$, $p_2 = \rho_2 R_2 T_2$ $\rightarrow \frac{p_1}{p_2} = \frac{\rho_1}{\rho_2} = \frac{1}{12}$

$\therefore \; p_2 = p_1 \times 12 = 14.7 \times 12 = 176.4 \text{ psia}$ $(R_1 = R_2)$

In the second case, $T_2 = 50°F \neq T_1$

$$M_1 = M_2 = \frac{V_1}{c_1} = \frac{V_2}{c_2} \rightarrow \frac{V_1}{V_2} = \sqrt{\frac{T_1}{T_2}}$$

$$Re_1 = Re_2 = \frac{V_1 L_1 \rho_1}{\mu_1} = \frac{V_2 L_2 \rho_2}{\mu_2} \rightarrow \frac{V_1}{V_2} \cdot 12 = \frac{\rho_2}{\rho_1} \cdot \frac{\mu_1}{\mu_2} = \sqrt{\frac{T_1}{T_2}} \cdot 12 \qquad (*)$$

From perfect gas law $\frac{\rho_2}{\rho_1} = \frac{p_2}{p_1} \cdot \frac{T_1}{T_2}$

Also, $\mu \sim T^{1/2}$. Substituting into eqn. (*): $\sqrt{\frac{T_1}{T_2}} \cdot 12 = \frac{p_2}{p_1} \frac{T_1}{T_2} \sqrt{\frac{T_1}{T_2}}$

$$\frac{p_2}{p_1} = 12 \times \frac{T_2}{T_1} = 12 \times \frac{(460+50)}{(460+25)} \Rightarrow p_2 = p_1 \times 12 \times 1.0515$$
$$= 14.7 \times 12 \times 1.0515 = 185.5 \text{ psia}$$

If the tunnel temp. changes, a different pressure is required.

6-7

A 1/20th scale model of a submarine is to be tested under both deep sea operating conditions and free surface submerging conditions. The operating speed of the prototype for both operations is 5 m/sec. It is noted that for deep water operations gravity effects are negligible compared to inertia effects.

a. Determine the significant parameters for each experiment.
b. Will a wind tunnel experiment on the model provide the value for the drag on the prototype under deep-sea operating conditions? Explain your answer. Assuming that the model and prototype are both operating at 10°C, under what conditions should the tunnel operate to avoid Mach number (compressibility) effects?
c. Will a wind tunnel experiment satisfy similarity conditions for the free surface operating conditions of the prototype? Explain your answer.
d. A towing basin experiment is to be performed to calculate the drag of the submarine under free surface operating conditions. Find the velocity of the model in the towing basin.

****

a) FOR THE DEEP SEA OPERATIONS THE SIGNIFICANT PARAMETERS ARE $Re = \frac{VL}{\nu}$ AND $C_D = \frac{D}{\frac{1}{2}\rho u^2 L^2}$

FOR FREE-SURFACE OPERATIONS THE SIGNIFICANT PARAMETERS ARE $Fr = \frac{V^2}{Lg}$ AND $C_D = \frac{D}{\frac{1}{2}\rho u^2 L^2}$

b) YES, DYNAMIC SIMILARITY IS SATISFIED BY REYNOLDS SIMILARITY IF COMPRESSIBILITY EFFECTS CAN BE REGULATED IN THE TUNNEL SO THEY ARE NEGLIGIBLE. THUS WE REQUIRE

$$(M)_m < 0.3 \quad \text{AND} \quad (Re)_m = (Re)_p$$

THUS

$$\frac{V_m}{a_m} < 0.3 \longrightarrow V_m < 0.3\,a_m = 0.3\sqrt{\gamma h T_m} = \sqrt{(1.4)(287)(283)}$$

OR

$$V_m < 101 \text{ m/sec}$$

THEN

$$\frac{\rho_m V_m L_m}{\mu_m} = \frac{\rho_p V_p L_p}{\mu_p}$$

OR

$$\rho_m = \rho_p \frac{V_p}{V_m} \frac{L_p}{L_m} \frac{\mu_m}{\mu_p}$$

OR

$$\rho_m > 1000 \frac{Kg}{m^3} \left(\frac{5}{101}\right)\left(\frac{20}{1}\right)\left(\frac{1.75 \times 10^{-5}}{1.307 \times 10^{-3}}\right)$$

$$\rho_m > 13.3 \ Kg/m^3$$

THEN

$$P_M = \rho_m R T_m \quad \text{OR} \quad P_M > 13.3 \frac{Kg}{m^3} \left(287 \frac{m^2}{s^2 K}\right)(283 \ K)$$

$$P_M > 1.08 \times 10^6 \ Pa$$

∴ DYNAMIC SIMILARITY CAN BE SATISFIED IN TUNNEL IF IT IS PRESSURIZED TO A PRESSURE GREATER THAN 1.08×10^6 Pa.

c) NO, THERE IS NO SURFACE IN THE WIND TUNNEL. THEREFORE GEOMETRIC SIMILARITY WOULD NOT EVEN BE SATISFIED.

d) FOR FREE SURFACE OPERATION THE DOMINANT PARAMETER IS THE FROUDE NUMBER. THEREFORE WE REQUIRE

$$(Fr)_m = (Fr)_p$$

OR

$$\left(\frac{V^2}{Lg}\right)_m = \left(\frac{V^2}{Lg}\right)_p$$

$$V_m = V_p \sqrt{\frac{L_m}{L_p}} = 5 \ m/sec \sqrt{1/20} = 1.12 \ m/sec$$

TURBOMACHINERY'S DIMENSION ANALYSIS

6-8

A model centrifugal pump was operated at an angular speed of 189 rad/s and delivered 0.125 m^3/s of water with the given impeller design. If the prototype pump is to operate at an angular speed of 21 rad/s, determine the discharge of the prototype pump with a prototype to model scale ratio of 9:1.

****

The discharges of prototype and model pumps are related as:

$$\left(\frac{Q}{nD^3}\right)_p = \left(\frac{Q}{nD^3}\right)_m$$

Here, $Q_m = 0.125 \; m^3/s$, $n_m = 189$ rad/s

$n_p = 21$ rad/s $\quad \dfrac{D_p}{D_m} = \dfrac{9}{1}$

$$\text{Therefore, } Q_p = \frac{(nD^3)_p Q_m}{(nD^3)_m} = \frac{21 \times (9)^3 \times 0.125}{189}$$

$$\text{or, } Q_p = 10.1 \; m^3/s$$

7 VISCOSITY

MEASUREMENT OF VISCOSITY

7-1

Calculate the kinematic viscosity of air and nitrogen at 60°F and 20 psia.

****

$\nu = \dfrac{\mu}{\rho}$ $\therefore$, we need to know kinematic viscosity and density.

We can obtain μ from tables for air and nitrogen (assuming that they are not influenced by pressure changes):

For air $\mu = 3.7466 \times 10^{-7}$ lb-sec/ft^2

For nitrogen $\mu = 3.6340 \times 10^{-7}$ lb-sec/ft^2

Density at $T = 60°F$ and $p = 50$ psia

for air $\rho = \dfrac{P}{RT} = \dfrac{20 \times (144) \times 28.97}{1545.3 \times (60+460)} = 0.10383 \dfrac{\text{lbm}}{\text{ft}^3} = 0.0032276 \dfrac{\text{slug}}{\text{ft}^3}$

for nitrogen $\rho = \dfrac{P}{RT} = \dfrac{20 \times (144) \times 28.016}{1545.3 \times (60+460)} = 0.10041 \dfrac{\text{lbm}}{\text{ft}^3} = 0.0031213 \dfrac{\text{slug}}{\text{ft}^3}$

$\therefore$ for air $\nu = \dfrac{\mu}{\rho} = \dfrac{3.7466 \times 10^{-7}}{3.2276 \times 10^{-3}} = 1.1602 \times 10^{-4}$ ft^2/sec

for nitrogen $\nu = \dfrac{\mu}{\rho} = \dfrac{3.634 \times 10^{-7}}{3.1213 \times 10^{-3}} = 1.16427 \times 10^{-4}$ ft^2/sec

7-2

A flat plate slides down an inclined plane as shown at a speed of 18 ft/s. The inclined plane is lubricated by a thin film of oil having a viscosity of 0.002 lb_f-s/ft^2. Determine the thickness of the oil film. Assume the area of the plate in contact with the plane to be 9 ft^2 and its mass to be 2 slug.

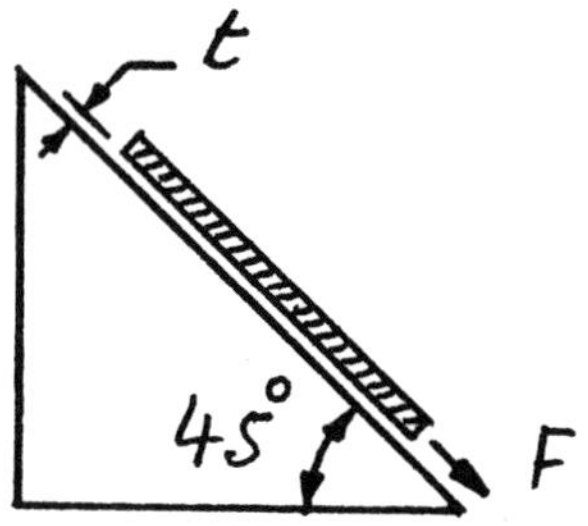

****

Force causing downward motion $= F = mg \sin 45°$

Newton's Law of viscosity: $F = \mu \frac{du}{dy} A$

Substituting values of different parameters in the above equation:

$$mg \sin 45° = 0.002 \times \frac{18}{t} \times 9$$

or, $t = \dfrac{0.002 \times 18 \times 9}{2 \times 32.174 \times 0.707} = 0.007121$ ft

or, $t = 0.085$ in.

7-3

Determine the viscosity of an unknown oil at 50°C utilizing a Saybolt Universal viscometer. The specific gravity of the oil is 0.86. The viscometer has the following dimensions: tube diameter - 1.76 mm, tube length - 12.25 mm, and headloss - 125 mm. It takes 92 seconds to collect a 60 ml sample.

****

$$\rho = S.G. \times \rho_{H_2O} = 0.86 \times 988.0 \ kg/m^3 = 849.7 \ Kg/m^3$$

$$g = 9.81 m/s^2 \qquad d = 0.00176 \ m \qquad h_L = 0.125 \ m \qquad l = 0.01225 \ m$$

$$V = 60 ml = 0.000\,060 \ m^3 \qquad t = 92 \ sec.$$

$$\mu = \left(\frac{\pi d^4 h_L}{128 V l}\right) \rho g t = \left(\frac{(3.14)(0.00176)^4 (0.125)}{(128)(0.000\,060)(0.01225)}\right)(849.7)(9.81)(92)$$

$$\mu = 0.0307 \ \frac{Kg \cdot m}{s^2} \frac{1}{m^2} s = 0.0307 \ N/m^2 \cdot s = 0.0307 \ Pa \cdot s$$

7-4

A capillary tube(10ft. long, .0794cm radius) is used to measure the viscosity of p-xylene. If the total pressure drop is .400 psi and a volumetric flowrate of .220 cm^3/s is observed, calculate the viscosity of p-xylene. The temperature is 20°C.

****

Start with the Hagen-Poiseuille Law:

$$Q = \frac{\left(\frac{\Delta P}{L}\right)\pi R^4}{8\mu} \quad \text{with} \quad R = .0794 \text{ cm.} \quad L = 10 \text{ ft.} = 305 \text{ cm.}$$

$$Q = .220 \ cm^3/s \qquad \Delta P = .400 \ psi = 2.76 \cdot 10^4 \ dyn/cm^2$$

Solving for viscosity and substituting,

$$\mu = \frac{\left(\frac{\Delta P}{L}\right)\pi R^4}{8Q} = \frac{(2.76 \cdot 10^4 dyn/cm^2)\pi(.0794 cm)^4}{8(305 cm.)(.220 \ cm^3/s)} = 6.42 \cdot 10^{-3} dyn/cm^2 s \ (\text{or } g\,cm^{-1} s^{-1})$$

$$\mu = .642 \ cp. \ \text{at } 20°C.$$

7-5

A cylindrical weight of 5 pounds falls at constant velocity inside a cylinder of 6-inch diameter. The diameter of the weight is 5.995 inches as shown in the diagram below. The oil film is crude oil at 150 degrees F. Determine the velocity of the weight.

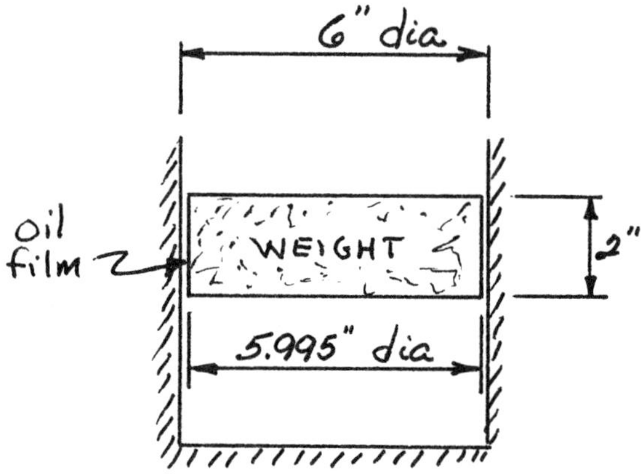

****

$$\tau = \mu \frac{dv}{dy} \Rightarrow dv = \tau \times dy \times \frac{1}{\mu}$$

$\mu = 7 \times 10^{-5} \frac{\text{lb-sec}}{\text{ft}^2}$ (obtained from viscosity table).

$$dy = \frac{6'' - 5.995''}{2} = .0025 \text{ in.}$$

$$\tau = \frac{F}{A} = \frac{5 \text{ lb}}{2''(\pi \times 6'')} = 0.13263 \text{ lb/in}^2$$

$$\therefore dv = .13263 \frac{\text{lb}}{\text{in}^2} (.0025 \text{ in}) \times \frac{1}{7 \times 10^{-5} \frac{\text{lb-sec}}{\text{ft}^2}} \times \frac{144 \text{ in}^2}{\text{ft}^2}$$

$$\underline{\underline{dv = 682.1 \text{ in/sec}}}$$

NOTE: dv is the velocity of the oil adhering to the weight relative to the velocity of the oil adhering to the cylinder wall. Since the latter velocity is zero, then dv represents the velocity of the weighT.

7-6

A certain oil is placed in the rotating drum viscometer. The motor develops a speed of 200 rpm. The force F'required to keep in place the stationary drum is 0.2 lb. The drag at the bottom is neglegible. The temperature is kept constant. What is the viscosity of this fluid?

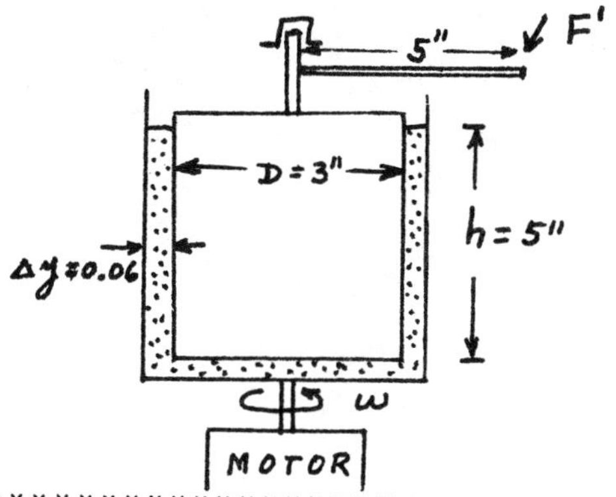

****

$$\mu = \tau / (\Delta v / \Delta y) \qquad \tau = \frac{F}{A}$$

$$A = \pi D * h \qquad F' * 5'' = F * (1.5'' + 0.03'')$$

$$= \pi * 3'' * 5'' \qquad 0.2\,lb * 5'' = 1.53''\,F$$

$$= 47.12\ in^2 \qquad F = 0.654\ lb$$

$$vel = 200\ \frac{rev}{min} * \frac{\pi D}{1\ rev}$$

$$= \pi 200 * 3''/min = 1{,}885''/min$$

$$\mu = \frac{0.654\ lb}{47.12\ in^2} \Big/ \left(\frac{1885\ in/min}{0.06\ in}\right) = 4.42 \times 10^{-7}\ \frac{lb\ min}{in^2}$$

$$= 4.42 \times 10^{-7}\ \frac{lb\ min}{in^2} \times \frac{60\ sec}{1\ min} * \frac{144\ in^2}{1\ ft^2} = 3.8 \times 10^{-3}\ \frac{lb\ sec}{ft^2}$$

TEMPERATURE EFFECTS

7-7

A 3m by 5m plate is moving through water at 70°C with a constant velocity of 100 m/sec. Calculate the total force on the plate due to the shear stress from the fluid. Assume a linear velocity profile.

****

$u(y) = Ay + B$

NO SLIP CONDITIONS: $u(0) = 0$

$u(1m) = 100\ m/sec$

$u(0) = 0 \rightarrow 0 = B$

$u(1m) = 100\ FT/SEC \rightarrow A = 100\ SEC^{-1}$

THE SHEAR STRESS ON THE PLATE IS GIVEN BY

$$\tau = \mu \frac{du}{dy} = \mu A$$

FOR WATER AT 70°C

$$\ln \frac{\mu}{\mu_0} = a + b\left(\frac{T_0}{T}\right) + c\left(\frac{T_0}{T}\right)^2$$

$T_0 = 273.16°K = 0°C$

$\mu_0 = 0.001792\ \frac{Kg}{m\text{-}sec}$

$a = -1.94$, $b = -4.80$, $c = 6.74$

$$\ln \frac{\mu}{\mu_0} = -1.94 - 4.80\left(\frac{273.16}{70 + 273.16}\right) + 6.74\left(\frac{273.16}{70 + 273.16}\right)^2$$

$$= -1.9379$$

$$\mu = 0.000258\ \frac{Kg}{m\text{-}sec}$$

THEN

$$\tau = \left(0.000258\ \frac{Kg}{m\text{-}sec}\right)\left(100\ sec^{-1}\right) = 0.00258\ \frac{Kg}{m\text{-}sec^2}$$

THE TOTAL FORCE ON THE PLATE IS

$$F = \tau A = \left(0.00258\ \frac{N}{m^2}\right)\left(15\ m^2\right) = 0.0387\ N$$

REYNOLD'S NUMBER

7-8

A 3" ID smooth pipe 100 ft. long carries 100 gpm of crude oil (SG=0.86). Determine the Reynold's number when the oil is:
a) 80° F
b) 115° F

****

$$Q = \frac{100 \text{ gpm}}{449 \text{ gpm/cfs}} = .223 \text{ cfs}$$

$$Q = AV; \quad V = \frac{Q}{A} = \frac{.223 \text{ ft}^3/\text{s}}{\frac{\pi}{4}\left(\frac{3}{12}\right)^2} = 4.543 \text{ ft/s}$$

$$\rho = .86\left(1.94 \text{ slug/ft}^3\right) = 1.67 \text{ slug/ft}^3$$

$$\text{At } 80°\text{F}, \ \mu = 1.5 \times 10^{-4} \text{ Lb}\cdot\text{s/ft}^2$$

$$\text{At } 115°\text{F}, \ \mu = 1.0 \times 10^{-4} \text{ Lb-s/ft}^2$$

From Dynamic Viscosity Vs. Temp. graph (Ans. may vary)

$$R = \frac{VD\rho}{\mu}; \quad R_{80} = \frac{(4.543)(.25)(1.67)}{1.5 \times 10^{-4}} = 12{,}600 \text{ ANS}$$

$$R_{115} = \frac{(4.543)(.25)(1.67)}{1.0 \times 10^{-4}} = 19{,}000 \text{ ANS}$$

7-9

Air flows in a pipe of 2 ft diameter at 80°F with a velocity of 10 ft/s. Determine whether this flow is laminar or turbulent. In order to have the same Reynolds number in a pipe of 1 ft diameter, what should the velocity be ?

****

For 80°F , $\nu = 1.69 \times 10^{-4}$ ft²/s

Reynolds number $\quad Re = \dfrac{U \cdot D}{\nu} = \dfrac{10 \times 2}{1.69 \times 10^{-4}} = 118345$

$Re_D = 118345 > 2300$: This flow is turbulent.

Velocity for 1 ft diameter pipe :

$$Re_D = 118345 = \frac{U \cdot D}{1.69 \times 10^{-4}} \quad \rightarrow \quad U = 20 \text{ ft/s}$$

or $\quad \left(\dfrac{U \cdot D}{\nu}\right)_1 = \left(\dfrac{U \cdot D}{\nu}\right)_2 \quad \rightarrow \quad (U \cdot D)_1 = (U \cdot D)_2$

$$10 \times 20 = U \times 1 \quad \rightarrow \quad U = 20 \text{ ft/s}.$$

7-10

Crude oil is to be transported in a 0.78 meter diameter pipeline at a velocity of two meters per second. It will be kept at a temperature of 35°C. Calculate the Reynolds Number of the fluid.

****

$$\mathbb{R} = \frac{Vd\rho}{\mu}$$

$$Pa = \frac{kg}{m \cdot s^2}$$

SPECIFIC GRAVITY = 0.86

DENSITY OF WATER = 993.95 kg/m³

DENSITY OF CRUDE OIL = 993.95 × .86 = 854.8 kg/m³

VISCOSITY OF CRUDE OIL = 5.4×10^{-3} Pa·s

$$\mathbb{R} = \frac{(2\ m/s)(0.78\ m)(854.8\ kg/m^3)}{(5.4 \times 10^{-3}\ Pa \cdot s)} = \underline{\underline{2.47 \times 10^5}}$$

8

SURFACE RESISTANCE AND BOUNDARY LAYER

LAMINAR AND TURBULENT FLOWS

8-1

Water at 75°C is flowing in a 1 inch schedule 80 steel pipe. The flow rate is 300 L/Min. Determine the Reynolds No. and if flow is laminar or turbulent.

****

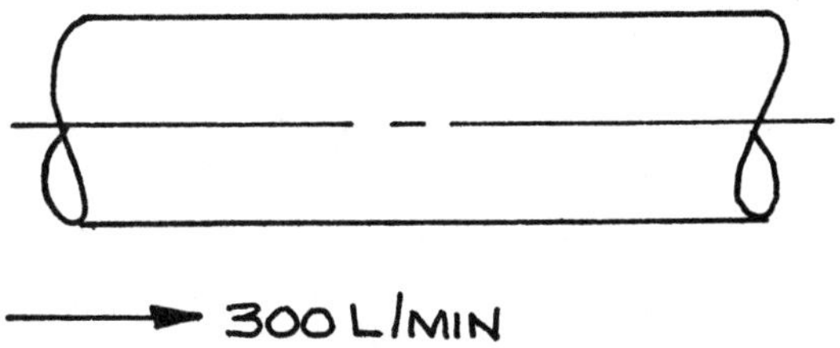

$D_i = 0.0243 \text{ M}$

$A_i = 4.636 \cdot 10^{-4} \text{ M}^2$

$\nu = 3.83 \cdot 10^{-7} \text{ M}^2/\text{s}$

$$V = Q/A = \frac{300 \cdot 10^{-3}}{(60)\, 4.636 \cdot 10^{-4}}$$

$V = 10.785 \text{ M/s} = 35.4 \text{ FPS}$

$$R_N = \frac{VD}{\nu} = 68.43 \cdot 10^4$$

$R_N = 0.684 \cdot 10^6$ TURBULENT FLOW

8-2

Oil flows in a pipe of 0.2 m diameter with a mean velocity of 1 m/s. Oil temperature is 20°C. Assume fully developed flow and calculate the skin friction coefficient, wall shear stress, pressure gradient and pressure loss per unit length of the pipe.

****

For oil at 20°C, $\nu = 901 \times 10^{-6}\ m^2/s$, $\rho = 888\ kg/m^3$

Reynolds number $Re_D = \dfrac{U_o \cdot D}{\nu} = \dfrac{1 \times 0.2}{901 \times 10^{-6}} = 221.98 < 2300$

Therefore, flow in the pipe is laminar. For fully developed laminar flow, skin friction is given by $C_f = \dfrac{16}{Re_D}$; $C_f = \dfrac{16}{221.98} = 0.07208$

Pressure drop per unit length :

$$\frac{\Delta P}{L} = \frac{4}{D} \cdot C_f \cdot \frac{1}{2} \rho U_o^2 = \frac{4}{0.2} \times 0.07208 \times \frac{1}{2} \times 888 \times 1^2 = 640\ \frac{Pa}{m}$$

Maximum velocity on the centerline $U_{max} = 2U_o = 2 \times 1 = 2\ m/s$

Therefore, wall shear stress :

$$\tau_w = \frac{2 \mu U_{max}}{r_o} = \frac{2 \times (800 \times 10^{-3}) \times 2}{0.1} = 32\ Pa/m^2$$

or : $\tau_w = C_f \cdot \frac{1}{2} \rho U_o^2 = 0.07208 \times \frac{1}{2} \times 888 \times 1^2 = 32\ Pa/m^2$

Pressure gradient :

$$\frac{dP}{dx} = -2\frac{\tau_w}{r_o} = -2 \times \frac{32}{0.1} = -640\ Pa/m.$$

Since the flow is fully developed, this is equal to $\Delta P/L$.

Or, pressure gradient can be calculated in a different way :

$$\frac{dP}{dx} = -\frac{8 \mu U_o}{r_o^2} = -\frac{8 \times (800 \times 10^{-3}) \times 1}{(0.1)^2} = -640\ Pa/m$$

8-3

A knuckleball is a type of baseball pitch that is thrown without any spin on the ball. The seams cause flow transitions over the spherical surface so that the baseball(diameter D=3 in.) moves in an unpredictable manner in response to varying drag and lift. Estimate the best speed(in miles per hour) to throw a knuckleball on a calm summer evening(T=60°F.).

****

Transition from laminar to turbulent flow generally depends on the Reynold's Number for the system of interest. In the case of a sphere, the Reynold's Number is:

$$Re = \frac{\rho \langle v \rangle D}{\mu} = \frac{\langle v \rangle D}{\nu}$$ where $\langle v \rangle$ is the average velocity.

The desired velocity is determined by the value of the Reynold's Number for the flow transition Re_{trans}. From experimental results for a smooth sphere, the onset of turbulence and a sharp drop in C_D occur at $Re_{trans} \approx 2 \cdot 10^5$. From tables of physical properties for air, the kinematic viscosity ν is $1.59 \cdot 10^{-4}$ ft²/s at 60°F.

Hence,

$$\langle v \rangle = \frac{\nu \, Re_{trans}}{D} = \frac{(1.59 \cdot 10^{-4} \text{ ft}^2/\text{s})(2 \cdot 10^5)}{(.25 \text{ ft})} = 130 \text{ ft/s or } 87 \text{ mph}$$

In practice, the threads cause an early onset of turbulence at a lower Reynold's Number. (A Knuckleball is considered to be a slow pitch in baseball!)

8-4

Water at a temperature of 100°C is flowing through a tube of 20 mm diameter with a velocity of 10 mm/s. Determine the friction factor of the tube.

****

The Reynolds number $= R = \dfrac{\rho V D}{\mu}$

From the tables of Fluid Properties at 100°C, density of water, $\rho = 958.380$ kg/m^3, and viscosity, $\mu = 0.000284$ Pa s. Then

$$R = \frac{958.380 \times 0.010 \times 0.020}{0.000284}$$

or, $R = 674.92$

Friction factor, $f = \dfrac{64}{R} = \dfrac{64}{674.92} = 0.095$

8-5

Water at 70°F flows over a flat plate of 6ft length with a velocity of 1 ft/s. Determine the velocity at a point at a distance of 3 ft from the leading edge and 0.01 ft from the surface of the plate.

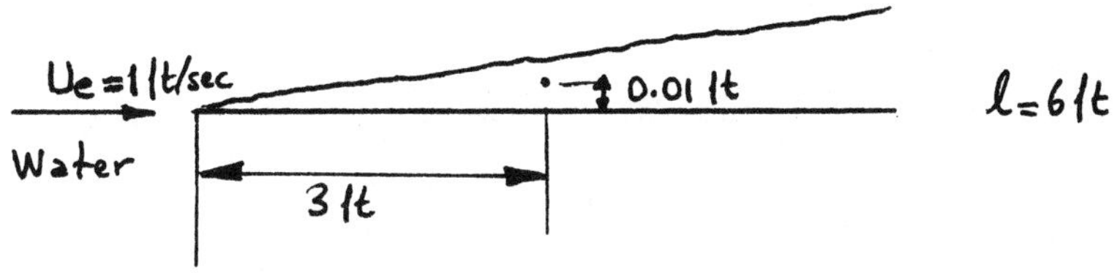

****

For water, $\nu = 1.06 \times 10^{-5}$ ft²/sec at 70°F.

Therefore, Reynolds number $Re_x = \frac{U_e \cdot x}{\nu} = \frac{1.0 \times 3}{1.06 \times 10^{-5}} \hat{=} 283020$

This value is smaller than 5×10^5 where transition occurs.

Therefore, laminar flow exists at $x = 3$ ft. For laminar boundary layer on a flat plate, the thickness of the boundary layer is expressed by $\delta = x \cdot \frac{5}{Re_x^{0.5}} = 3 \cdot \frac{5}{532} = 0.0282$ ft.

The point we are considering is at 0.01 ft < 0.0282 ft, and in the laminar boundary layer:

$$\frac{y}{x} \cdot Re^{0.5} = \frac{0.01}{3} \times 532 = 1.773$$

Using Blasius solution, we can read $U/U_e = 0.58$.

Therefore, $U = 0.58 \times U_e = 1 \times 0.58 = 0.58$ ft/sec.

DARCY'S EQUATION

8-6

Oil flows through an inclined pipe as shown. Flow conditions are given at two points in the flow. Assume laminar flow.

a. Determine the direction of the flow.
b. Determine the flow rate.
c. Calculate the friction factor.
d. Is the flow really laminar ?

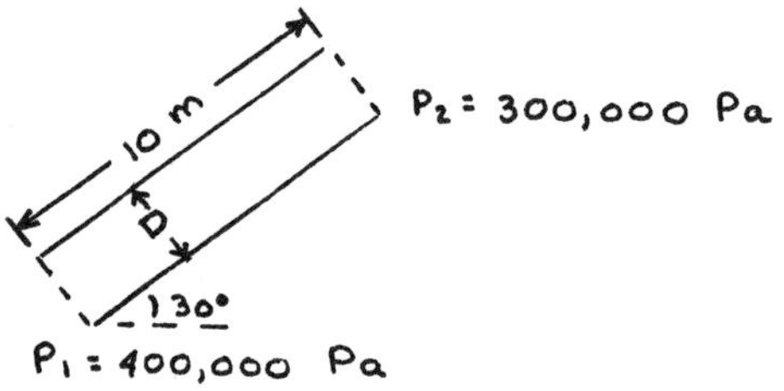

$D = 0.04$ m
$\rho = 900$ Kg/m²
$\nu = 2 \times 10^{-4}$ m²/sec

****

a) ASSUME FLOW FROM 1 TO 2. THEN BERNOULLIS EQUATION WITH HEAD LOSS TERMS YIELD

$$\frac{P_1}{\rho g} + \frac{V_1^2}{2g} + z_1 = \frac{P_2}{\rho g} + \frac{V_2^2}{2g} + z_2 + h_f$$

CONTINUITY YIELDS

$$(\rho A V)_1 = (\rho A V)_2 \longrightarrow V_1 = V_2$$

TAKE THE DATUM AT 1. THEN

$$z_1 = 0 \qquad z_2 = (10\text{ m})(\sin 30°) = 5\text{ m}$$

THUS WE HAVE

$$h_f = \frac{P_1 - P_2}{\rho g} - z_2 = \frac{(400000 - 300000)\text{ Pa}}{(900\text{ Kg/m}^3)(9.81\text{ m/sec}^2)} - 5\text{m} = 6.33\text{ m}$$

SINCE $h_f > 0$ THE FLOW IS IN THE DIRECTION ASSUMED.

b) FOR LAMINAR FLOW

$$h_f = \frac{64}{Re}\,\frac{L}{D}\,\frac{V^2}{2g} \qquad \text{WHERE} \qquad Re = \frac{VD}{\nu}$$

THUS

$$h_f = \frac{64}{\left(\frac{VD}{\nu}\right)}\,\frac{L}{D}\,\frac{V^2}{2g} = \frac{32\nu LV}{gD^2}$$

OR

$$V = \frac{gD^2 h_f}{32\nu L} = \frac{(9.81\ m/sec^2)(6.33\ m)(0.04\ m)^2}{32(2\times10^{-4}\ m^2/sec)(10\ m)} = 1.552\ m/sec$$

THE FLOW RATE IS GIVEN BY

$$Q = VA = (1.552\ m/sec)(\pi)(0.02\ m)^2 = 1.95\times10^{-3}\ m^3/sec$$

c)

$$f = \frac{64}{Re} \qquad Re = \frac{VD}{\nu} = \frac{(1.552\ m/sec)(0.04\ m)}{2\times10^{-4}\ m^2/sec} = 310.4$$

$$f = 0.206$$

d) $Re < 2300 \longrightarrow$ FLOW IS LAMINAR

MOODY DIAGRAM AND ROUGHNESS

8-7

A 200 mm diameter pipe with a roughness projection of 2 mm carries a flow of 0.2 cubic meters per sec of water. Calculate the head loss in 100 meters of this pipe.

****

RELATIVE ROUGHNESS $- \frac{e}{d} = \frac{2\ mm}{200\ mm} = 0.01$

$V = \frac{Q}{A} = \frac{0.2\ m^3/s}{\pi (0.1)^2} = 6.37\ m/s$ $\qquad d = 0.2\ m$

DENSITY $- \rho = 998.2\ kg/m^3$ @ $20°C$

VISCOSITY $- \mu = 10 \times 10^{-4}\ Pa \cdot s$ @ $20°C$

REYNOLD'S NUMBER $- R = \frac{V d \rho}{\mu} = \frac{(6.37)(0.2)(998.2)}{(10 \times 10^{-4})} = 1.27 \times 10^6$

DETERMINE FRICTION FACTOR USING MOODY DIAGRAM

$f = \underline{\underline{0.0377}}$ @ $\frac{e}{d} = 0.01$ AND $R = 1.27 \times 10^6$

CALCULATE HEADLOSS IN PIPE $- h_L = f \frac{l}{d} \frac{V^2}{2g}$

$h_l = (0.0377)\left(\frac{100\ m}{0.2\ m}\right) \frac{(6.37\ m/s)^2}{(2)(9.81\ m/s^2)} = 38.98 = \underline{\underline{39.0\ m}}$

8-8

Water at 60° F is flowing in a 2" ID pipe at 5 ft/sec. ϵ is approximately 0.0002 ft. What pressure drop, in psi, would you expect in 500 ft. of pipe? What horsepower pump is needed to overcome this head loss?

****

$\gamma = 62.4 \text{ Lb/ft}^3$
$\rho = 1.94 \text{ slug/ft}^3$
Kinematic viscosity $= \nu = 1.21 \times 10^{-5} \frac{\text{ft}^2}{\text{sec}}$

} Table: properties of water

Reynold's Number $= N_R = \frac{VD}{\nu} = \frac{(5 \text{ ft/s})(\frac{2}{12} \text{ ft})}{1.21 \times 10^{-5} \text{ ft}^2/\text{s}} = 68{,}870$

Relative Roughness $= \frac{D}{\epsilon} = \frac{\frac{2}{12}}{.0002} = 833$

$f = .023$ (MOODY'S DIAGRAM)

DARCY'S EQN : $h_L = \frac{fLV^2}{D2g} = \frac{.023(500)(5)^2}{(\frac{2}{12})(2 \times 32.2)} = 26.8'$

Assume pipe is horizontal, uniform diameter, so that: $\frac{P_2 - P_1}{\gamma} = h_L$

$\Delta P = (26.8 \text{ ft})\left(62.4 \frac{\text{Lb}}{\text{ft}^3}\right)\frac{\text{ft}^2}{144 \text{ in}^2} = \boxed{11.6 \text{ psi}}$ ANS

$E_p = h\gamma Q$; $Q = AV = \frac{\pi(\frac{2}{12})^2}{4}(5) = .109 \frac{\text{ft}^3}{\text{s}}$

$= \frac{(26.8 \text{ ft})(62.4 \frac{\text{Lb}}{\text{ft}^3})(.109 \frac{\text{ft}^3}{\text{s}})}{550 \frac{\text{ft-Lb}}{\text{sec}} / \text{HP}} = \boxed{.332 \text{ HP}}$ ANS

8-9

Water of kinematic viscosity 1.217×10^{-5} ft^2/sec is pumped from reservoir 1 to reservoir 2 at a constant flow rate of 5 ft^3/sec. The surface of reservoir 2 is under a pressure of 40 psig and is 300 ft above the surface of reservoir 1. The pump delivers 250 hp to the fluid. The pipe is made of galvanized iron. What size pipe is required for this operation? List all assumptions.

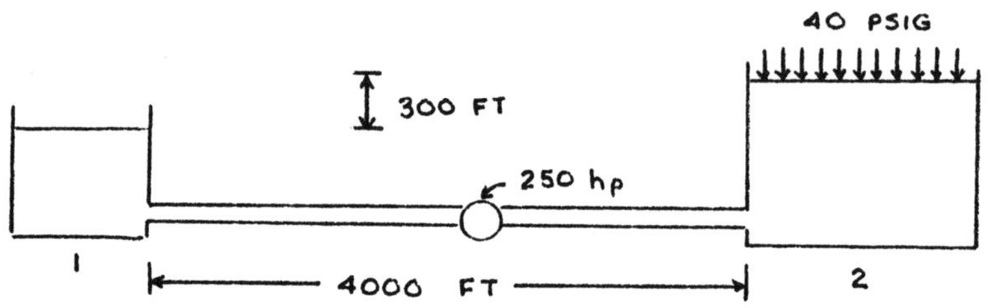

****

THE ASSUMPTIONS FOR THIS PROBLEM ARE

1) THE FLOW IS STEADY
2) THE FLOW IS INCOMPRESSIBLE
3) WATER IS A NEWTONIAN FLUID
4) ENTRY LOSSES TO PIPE ARE NEGLIGIBLE
5) NO MINOR LOSSES OCCUR
6) THE PROCESS IS ADIABATIC
7) $\Delta u \simeq 0$
8) NO EXTERNAL ENERGY IS ADDED
9) THE PRESSURE AT THE SURFACE OF RESERVOIR IS APPROXIMATELY P_{ATM}
10) THE RESERVOIRS ARE LARGE

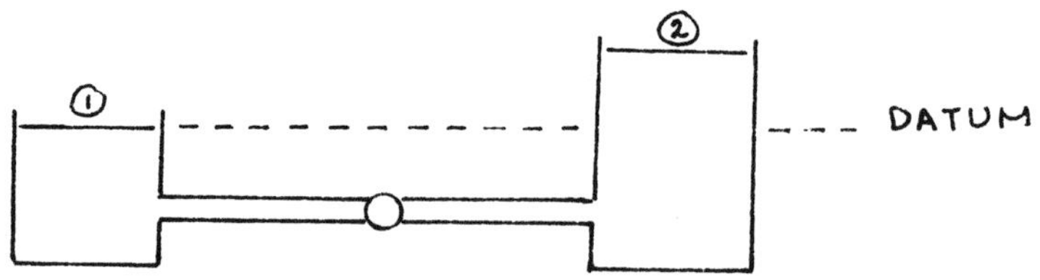

LET POINT ① BE AT THE SURFACE OF RESERVOIR ① AND POINT ② BE THE SURFACE OF RESERVOIR ②. WRITING THE ENERGY EQUATION BETWEEN ① + ② IN HEAD FORM

$$\frac{P_1}{\rho g} + \frac{V_1^2}{2g} + z_1 = \frac{P_2}{\rho g} + \frac{V_2^2}{2g} + z_2 + h_f + h_s$$

WHERE P_1 = PRESSURE (GAUGE) AT 1 = 0
P_2 = PRESSURE (GAUGE) AT 2 = 40 psig
V_1 = VELOCITY AT 1 = 0
V_2 = VELOCITY AT 2 = 0

$z_1 = 0$ FT

$z_2 = 300$ FT

h_s = PUMP HEAD $= -M$

h_f = HEAD LOSSES IN PIPE

$$M = \frac{(H_P)(550)}{\rho g Q} = \frac{(250\ hp)(550)}{(1.94)(32.2)(5)} = 440.2 \text{ FT}$$

THUS SOLVING FOR h_f

$$h_f = M - \frac{P_L}{\rho g} - z_2 = 440.2 \text{ FT} - \frac{(40\ psig)(144\ IN^2/FT^2)}{62.4\ LBS/FT^3} - 300 \text{ FT}$$

$$= 48 \text{ FT}$$

BUT

$$h_f = \frac{f v^2}{2g}\frac{L}{d}$$

WHERE f IS THE FRICTION FACTOR AND v IS THE VELOCITY THROUGH THE PIPE

$$v = \frac{Q}{A} = \frac{Q}{\pi d^2/4}$$

THUS

$$h_f = 2517\ f/d^5$$

HENCE

$$d^5 = 52.4\ f$$

AN ITERATION PROCEDURE WILL BE USED TO SOLVE THE ABOVE EQUATION. A VALUE OF f WILL BE ASSUMED AND d CALCULATED FROM THE ABOVE EQUATION. THEN A REYNOLDS NUMBER WILL BE CALCULATED.

$$Re = \frac{VD}{\nu} = \frac{4Q}{4d\nu} = \frac{5.23 \times 10^5}{d}$$

THE REYNOLDS NUMBER WILL BE USED TO DETERMINE A NEW FRICTION FACTOR FROM THE MOODY DIAGRAM. THE PROCESS CONTINUES UNTIL SUITABLE CONVERGENCE IS ACHIEVED.

GUESS $f = 0.02 \rightarrow d = 1.009$ FT $\rightarrow \varepsilon/d = 0.0005/1.009 \approx 0.0005$, $Re = 5.23 \times 10^5 \rightarrow f = 0.018 \rightarrow d = 0.988$ FT $\rightarrow \varepsilon/d = 0.0005$, $Re = 5.29 \times 10^5$

AT THIS POINT IT IS IMPOSSIBLE TO DISTINGUISH ANY DIFFERENCES ON THE MOODY DIAGRAM. THUS WE TAKE $d = 0.998$ FT. FOR PRACTICAL PURPOSES $d = 1$ FT.

8-10

When piping hot water (Between 130F and 180F) in steel pipe ($e = 4\times10^{-4}$ft) for diameters 2" and 4" and limiting the water velocity between 3ft/s and 10ft/s (a) What range of friction coefficient (f) is to be expected? (b) If you use f=0.024 for design, what is the maximum percentage of error in the loss head?

****

To enter the Moody diagram you need:

$$N_R = \frac{Dv}{\nu} \quad \text{and} \quad \text{Rel. Roughness} = \frac{D}{\varepsilon}$$

(Note that most Moody diagrams are based on $\frac{\varepsilon}{D}$ rather than $\frac{D}{\varepsilon}$)

the range of ν is 5.49×10^{-6} to $3.84\times10^{-6}\ \frac{ft}{s^2}$

the values for Rel. Roughness for 2" $\dfrac{2/12\ ft}{4\times10^{-4}ft} = 417$

for 4" $\dfrac{4/12\ ft}{4\times10^{-4}ft} = 833$

N_R for 2", 3 ft/s, 130°F $\quad N_R = \dfrac{1/6\times3}{5.49\times10^{-6}} = 0.9\times10^5$

same, 180°F $\quad N_R = \dfrac{1/6\times3}{3.84\times10^{-6}} = 1.3\times10^5$

N_R for 2", 10 ft/s, 130°F $\quad N_R = \dfrac{2/12\times10}{5.49\times10^{-6}} = 3.0\times10^5$

same, 180°F $\quad N_R = \dfrac{1/6\times10}{3.84\times10^{-6}} = 4.3\times10^5$

⋮

N_R for 4", 10 ft/s, 180 F $\quad N_R = \dfrac{4/12\times10}{3.84\times10^{-6}} = 8.7\times10^5$

from Moody's for 2" f = 0.0275 to 0.0260

for 4" f = 0.0215 to 0.0205

Total range 0.0205 to 0.0275

Maximum error 0.0240 − 0.0205 = 0.0035

Max % error $\dfrac{0.0035}{0.0205} = 17\%$

9 FLOW IN CONDUITS

FRICTION FACTOR AND LOSSES IN PIPES

9-1

Glycerin flows through a 1200 foot, 6 inch pipe with an average velocity of 20 ft/sec. Find the pressure drop and the pump power required.

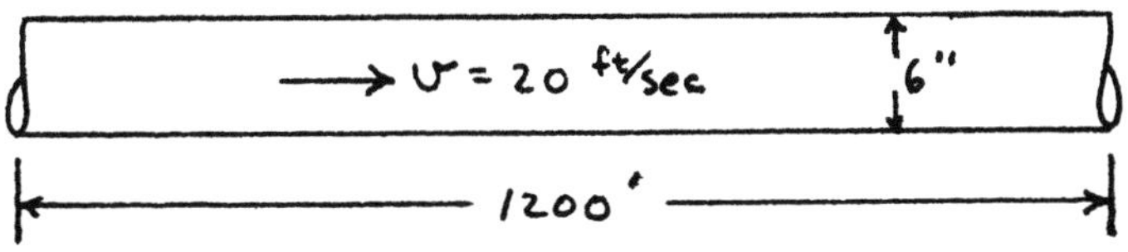

****

$$\text{Reynolds Number} = Re = \frac{\rho \upsilon D}{\mu} = \frac{6 \frac{lb_f\, sec^2}{ft^4}\; 20 \frac{ft}{sec}\; \frac{6}{12} ft}{.05 \frac{lb_f - sec}{ft^2}} = 1200$$

ρ and μ found from handbook

Since $Re < 2000$, flow is laminar

$$\therefore \text{ friction factor} = f = \frac{64}{Re} = \frac{64}{1200} = .0533$$

$$\Delta p = \rho g h_L = f \frac{L}{D} \frac{\rho \upsilon^2}{2} = .0533 \frac{1200}{6/12} \frac{6\,(20)^2}{2} = 1.54 \times 10^5 \frac{lb_f}{ft^2}$$

$$\dot{W}_s = Q \Delta p = \upsilon A \Delta p = \frac{20 \frac{ft}{sec} \frac{\pi}{4}\left(\frac{6}{12}\right) ft^2\; 1.54 \times 10^5 \frac{lb_f}{ft^2}}{550 \frac{ft - lb_f}{sec - hp}} = 1100 \text{ hp}$$

9-2

Water at 60°F flows in a smooth pipe of 1 ft diameter. If we consider a pipe length of 100 ft, determine the head loss for the velocities of 1 ft/s and 100 ft/s.

****

For 1 ft/sec : $Re_D = \frac{V.D}{\nu} = \frac{1 \times 1}{1.22 \times 10^{-5}} \cong 81967 > 2300$

i.e. flow is turbulent since $Re_D > 2300$.

Also,

$4000 < Re_D \cong 82000 < 10^5$, therefore we can use

$$f = \frac{0.316}{Re^{1/4}} = \frac{0.316}{81967^{0.25}} = 0.01867$$

Head loss: $h_f = \frac{f.L}{D} \cdot \frac{V^2}{2g_c} = \frac{0.01867 \times 100}{1} \times \frac{1^2}{2 \times 32.17} = 0.29$ ft

For 100 ft/sec : $Re_D \cong 8.2 \times 10^6$ flow is turbulent.

since $Re_D \geq 10^5$, we must use $f^{-\frac{1}{2}} = 2 \log Re . f^{\frac{1}{2}} - 0.8$

Assuming $f = 0.02$ (some close value) and for $Re_D = 8196700$,

we get after first iteration 0.00779
after second iteration 0.008387
after third iteration 0.008338
after fourth iteration $0.008342 = f$ close enough;

or, alternatively, for $Re_D = 8.2 \times 10^6$ and for smooth pipe from Moody diagram $f \cong 0.0083$ can be read.

Therefore,

$$h_f = \frac{0.0083 \times 100}{1} \times \frac{100^2}{2 \times 32.17} = 129 \text{ ft.}$$

9-3

Sea water is pumped in a horizontal cast iron pipe at a rate of 2 ft^3/s. The sea water is at a temperature of 50°F. Pipe diameter and length are 1 ft and 2000 ft, respectively. What are the head loss incurred and the power necessary to overcome this head loss ?

****

Velocity in the pipe $V = \frac{Q}{A} = \frac{Q}{\pi(\frac{D^2}{4})} = \frac{2 \times 4}{\pi \times 1^2} = 2.55 \text{ ft/sec}$

Reynolds number $Re_D = \frac{V.D}{\nu} = \frac{2.55 \times 1}{1.5 \times 10^{-5}} \cong 1.7 \times 10^5$

From Moody's relative roughness charts, we can read $\frac{k_s}{D} = 0.00085$ for 12-in. diameter cast iron pipe.

For this value of k_s/D and for the Reynold number, we can read $f = 0.021$ from Moody's resistance coefficient chart.

Therefore: head loss is

$$h_f = f.\frac{L}{D}.\frac{V^2}{2g_c} = 0.021 \times \frac{2000}{1} \times \frac{2.55^2}{2 \times 32.17} = 4.23 \text{ ft}$$

Necessary power :

$$P = \dot{w}.h_f = 2 \times 1.99 \times 4.23 \times 32.17 = 542 \frac{lb_f\text{-}ft}{sec}$$

$$= \frac{542}{550} = 0.985 \text{ hp}.$$

9-4

For a certain project, a water discharge of 8 cfs was required at a head loss of 25 ft of water. Recommend a commercially available pipe size along with its friction factor. The length of the pipe is 100 ft.

****

The friction factor formula for a pipe is:

$$\frac{D^5}{f} = \frac{8LQ^2}{\pi^2 g h_f} = \frac{8 \times 100 \times 64}{\pi^2 \times 32.174 \times 25}$$

$$\frac{D^5}{f} = 6.449$$

for $f = 0.02$, $D = 7.966$ in.

for $f = 0.021$, $D = 8.045$ in.

Therefore, a pipe of 8 in. diameter with a friction factor of 0.02 is recommended.

9-5

A utility company is using a concrete pipe for a water main. Determine the pressure drop and (f) for:

a) 12 inch inside diameter concrete pipe
b) 1.5 miles long
c) flow rate of 18 CFS, water at 60°F.

****

FLOW RATE:

WATER AT 60°F = 18 CFS

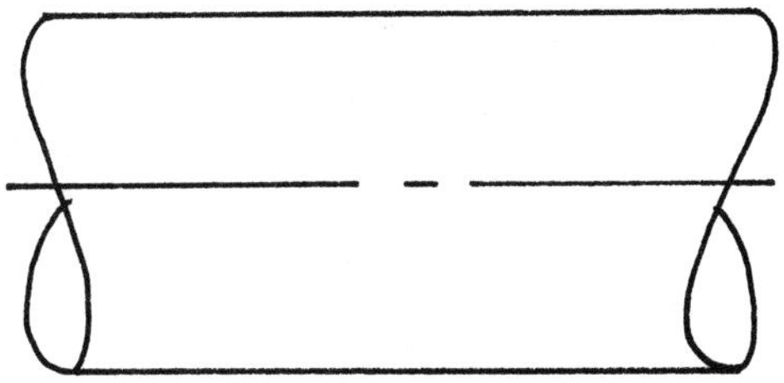

FIND REYNOLDS NO.

$R_N = \frac{VD}{\nu}$ $\quad V = Q/A$

$R_N = \frac{(22.9)(1)}{1.21 \cdot 10^{-5}} = 1.89 \cdot 10^{6}$

(ϵ) CONCRETE $= 4 \cdot 10^{-3}$ Ft. $\quad D/\epsilon = 250$

USE EQUATION FOR (f) $\Rightarrow$ $\frac{1}{\sqrt{f}} = 2 \log_{10}(3.7\ D/\epsilon)$

FRICTION FACTOR $\underline{(f) = 0.028}$

HEAD LOSS $H_L = f \frac{L}{D} \frac{V^2}{2g} = 1808$ Ft.

PRESSURE DROP $\Delta P = \gamma \Delta H_L$

$\Delta P = \frac{62.4\,(1808)}{144} = 783.7$ PSI

$\underline{\Delta P = 783.7 \text{ PSI}}$

9-6

Water at 60 F flows through a pipe with a head loss of 100 ft. in a length of 10,000 ft. The flow rate is 1.00 CFS. The material is concrete with f=0.023
What size pipe is required?

****

DARCY'S EQN.: $h_L = \frac{fLV^2}{D2g}$; $V = \frac{Q}{A} = \frac{Q}{\frac{\pi D^2}{4}} = \frac{4Q}{\pi D^2}$

$$h_L = \frac{fLV^2}{D2g} = f\frac{L}{D}\left(\frac{4Q}{\pi D^2}\right)^2 \frac{1}{2g} = \frac{8LQ^2 f}{\pi^2 g D^5}$$

$$D = \left(\frac{8LQ^2 f}{\pi^2 g\, h_L}\right)^{\frac{1}{5}}$$

$$= \left[\frac{8(10{,}000)(1)^2(.023)}{\pi^2(32.2)(100)}\right]^{\frac{1}{5}} = (.0579)^{\frac{1}{5}}$$

$$= .5656 \text{ ft}$$

$$= 6.79''$$

Use 8" or check with supplier to see what sizes are available.

9-7

A new 12 inch diameter commercial steel pipe carries 10 cubic feet per second. At point A the elevation is 400 ft. and at a point B, 1500 feet away, the elevation is 480 feet. If a pressure of 35 psi is to be maintained at point B, what is the pressure that is required at point A? Assume the water temperature to be 50°F.

****

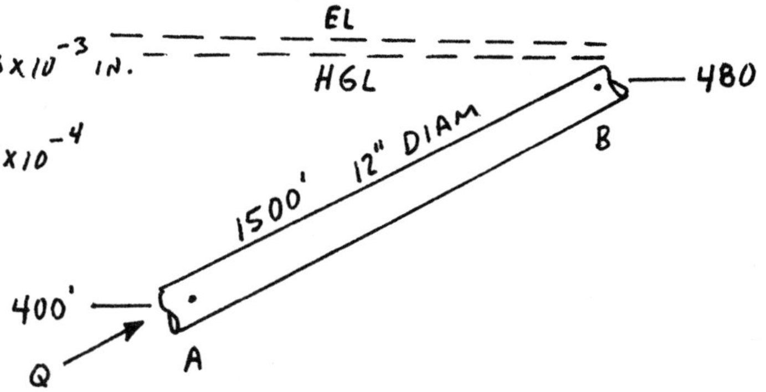

$e = 0.045\ mm = 1.773 \times 10^{-3}\ in.$

$\frac{e}{d} = \frac{1.773 \times 10^{-3}}{12} = 1.48 \times 10^{-4}$

$V = \frac{Q}{A} = \frac{10}{\pi(.5)^2} = 12.7\ fps$

$\nu = 1.410 \times 10^{-5}\ ft^2/s$

REYNOLDS NUMBER $= \frac{Vd}{\nu} = \frac{(12.7\ fps)(1\ ft)}{(1.410 \times 10^{-5}\ ft^2/s)} = 9.0 \times 10^{5}$

FROM THE MOODY DIAGRAM $f = 0.014$

HEADLOSS $- h_L = f\frac{l}{d}\frac{V^2}{2g} = (0.014)\left(\frac{1500}{1}\right)\frac{(12.7)^2}{(2)(32.2)} = \underline{\underline{52.6\ ft}}$

BERNOULLI'S EQUATION $- \frac{P_A}{\gamma} + \frac{V_A^2}{2g} + Z_A = \frac{P_B}{\gamma} + \frac{V_B^2}{2g} + Z_B + h_L$ (velocity heads cancelled)

$\frac{P_A}{62.41} + 400 = \frac{(35)(144)}{62.41} + 480 + 52.6$

$P_A = \underline{\underline{13{,}316\ psf}} = \underline{\underline{92.5\ psi}}$

9-8

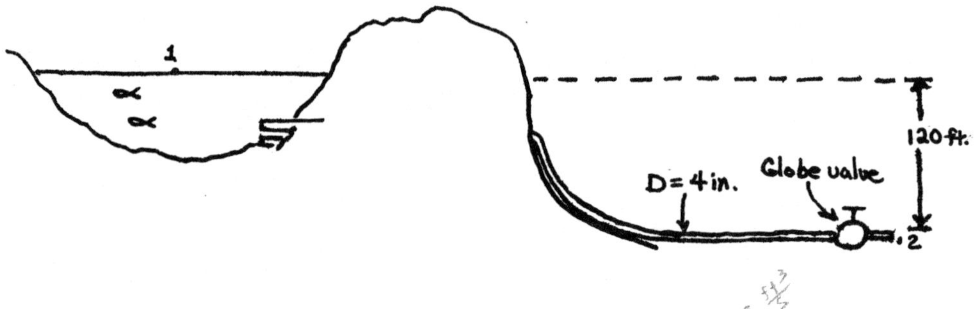

$.5568 \frac{ft^3}{s}$

$.0158 \frac{m^3}{s}$

A small mountain lake is to be used as a water source to irrigate a nearby field (see sketch). If the required delivery rate is 250 gal/min, will 4 in. diameter commercial steel pipe be sufficiently large for the job? The total length of the pipe is 2000 ft. You may neglect the loss effects of any bends. Represent minor losses with equivalent length where possible.

****

Assume steady, incompressible flow; $v_1 \approx 0$; $P_1 = P_2 = P_{atm}$

Use the head loss equation: $\left(\frac{P_1}{\rho} + \frac{v_1^2}{2} + gz_1\right) - \left(\frac{P_2}{\rho} + \frac{v_2^2}{2} + gz_2\right) = h_{LT} \ \frac{ft^2}{s^2}$

with $h_{LT} = h_L + h_{Lm}$.

$$g(z_1 - z_2) - \frac{v_2^2}{2} = h_L + h_{Lm} \qquad h_L = 2 f_f \frac{L}{D} v^2 \quad \text{where } \frac{L}{D} = \frac{2000\,ft}{.333\,ft.} = 6000$$

$$(h_{Lm})_{valve} = 2 f_f \frac{L_{eq}}{D} v^2 \qquad \left\{ \frac{L_{eq}}{D} = 350 \text{ for fully opened globe valve} \right\}$$

$$(h_{Lm})_{entrance} = K \frac{v^2}{2} \qquad \{ K = .34 \text{ for square edged entrance} \}$$

Substituting,

$$g(z_1 - z_2) - \frac{v^2}{2} = 2f \frac{L}{D} v^2 + 2f \frac{L_{eq}}{D} v^2 + K \frac{v^2}{2}$$

or

$$v = \left[\frac{g(z_1 - z_2)}{\frac{1}{2} + 2f\frac{L}{D} + 2f\frac{L_{eq}}{D} + \frac{K}{2}} \right]^{\frac{1}{2}}$$

For commercial steel pipe (4 in. diameter), $\frac{e}{D} = .00045$ or $D/e = 2200$. The friction factor from a graph is taken from the fully turbulent region as an initial guess. $f \approx .0042$

$$v = \left[\frac{(32.2\ ft/s^2)(120\ ft)}{\frac{1}{2} + 2(.0042)(6000) + 2(.0042)(350) + \frac{.34}{2}} \right]^{\frac{1}{2}} = 8.46\ ft/s$$

To improve the estimate, the Reynold's Number is calculated.

$$Re = \frac{\rho v D}{\mu} \qquad \text{Assume } T = 80°F., \text{ then } \rho = 62.2\ lbm/ft^3,\ \mu = 5.78 \cdot 10^{-4}\ \frac{lbm}{ft \cdot s}$$

$$Re = \frac{(62.2\ lbm/ft^3)(8.46\ ft/s)(.333\ ft)}{(5.78 \cdot 10^{-4}\ lbm/ft\ s)} = 3.03 \cdot 10^5$$

With $Re = 3 \cdot 10^5$ and $D/e = 2000$, $f = .0046 \Rightarrow v = 8.09\ ft/s$

$$Q = vA = v\frac{\pi D^2}{4} = \frac{(8.09\ ft/s)\pi(.333\ ft)^2}{4} = .706\ ft^3/s$$

$$Q = (.706\ ft^3/s)\left(\frac{7.48\ gal}{ft^3}\right)\left(\frac{60\ s}{min}\right) = 317\ gal/min$$ {So the pipe is more than large enough.}

MINOR LOSSES

9-9

Water if flowing between two reservoirs. Compute the flowrate in the pipes. Include minor losses. All fittings are flanged.

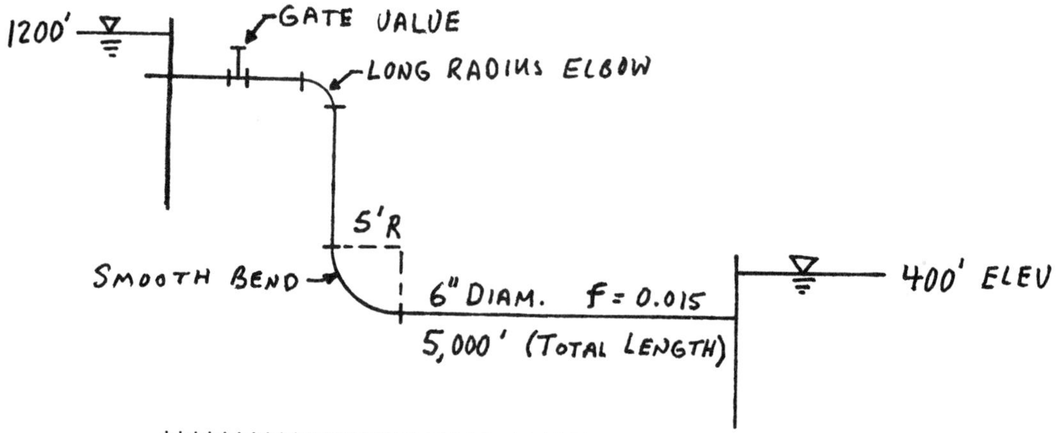

****

LOSS COEFFICIENTS – K_L

RE-ENTRANT ENTRANCE	– 0.8
GATE VALVE (WIDE OPEN)	– 0.1
LONG RADIUS ELBOW	– 0.2
PIPE EXIT	– 1.0
SMOOTH BEND ($R/d = 10$)	– 0.33

WRITE BERNOULLI'S EQUATION BETWEEN TWO RESERVOIRS WITH $P/\gamma = 0$ AND $\frac{V^2}{2g} = 0$

$$z_{1200} = z_{400} + h_L$$

$$\therefore\ h_L = 1200 - 400 = 800 \text{ ft.}$$

CALCULATE HEADLOSS EQUATION VALUES

$$h_L = (h_{L_{ENT.}} + h_{L_{VALVE}} + h_{L_{ELBOW}} + h_{L_{BEND}} + h_{L_{PIPE}} + h_{L_{EXIT}}) = 800 \text{ ft.}$$

$$= \left(K_{L_{ENT}} + K_{L_{VALVE}} + K_{L_{ELBOW}} + K_{L_{BEND}} + f\frac{l}{d} + K_{L_{EXIT}}\right)\frac{V^2}{2g}$$

$$= \left(0.8 + 0.1 + 0.2 + 0.33 + (0.015)\left(\frac{5000}{0.5}\right) + 1.0\right)\frac{V^2}{2(32.2)} = 800$$

$V = \underline{\underline{18.38 \text{ fps}}}$ $Q = AV = \pi(.25)^2(18.38) = \underline{\underline{3.6 \text{ cfs}}}$

9-10

Find the required pressure at point "A" in the tank in order to meet the following requirements:

a) Flow Rate = 435 L/Min
b) Water at 20°C
c) 2 in. Sch 40 pipe, 38 meters long
d) Standard elbow - other fittings shown, (steel)

****

WATER @ 20°C

$\nu = 1.02 \cdot 10^{-6}\ M^2/S$

$D_i = .0525\ M$

$A_i = 2.168\ 10^{-3}\ M^2$

$Q = .00725\ M^3/S$

$V = 3.344\ M/S$

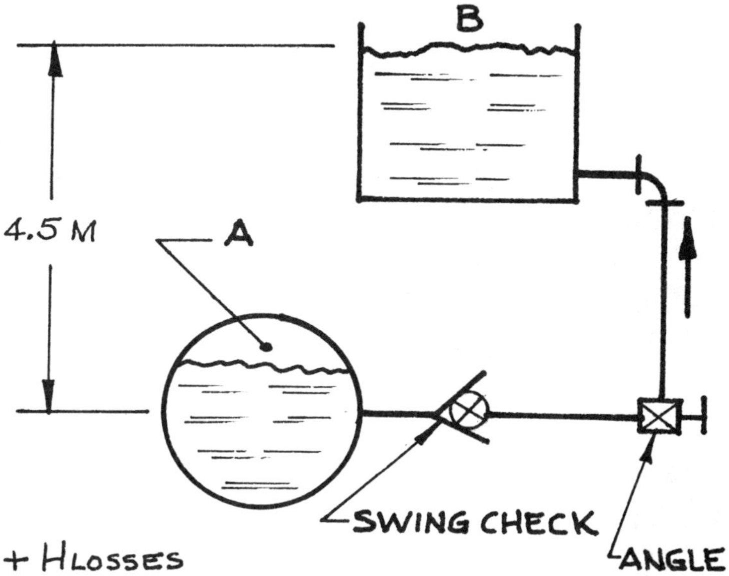

$$\frac{P_A}{\gamma} = \frac{P_B}{\gamma} + 4.5\,M + H_{LOSSES}$$

(P_B/γ struck out, = 0)

LOSSES:

ENTR	VALVE	VALVE	PIPE	EXIT	ELBOW
1	135	145	$f\frac{L}{D}$	1	30

REYNOLDS: $N_R = \frac{VD}{\nu} = 1.72 \cdot 10^5$ MOODY (f) $\cong .020$

$$H_L = \frac{(3.344)^2}{2g}\left[1 + .02(135) + .02(145) + \frac{.02(38)}{.0525} + 1 + .02(30)\right]$$

$$\frac{P_A}{\gamma} = 4.5 + 12.924 = 17.42\ M$$

$P_A = 17.42\,(9.79)$ $\underline{P_A = 170.6\ KPa}$

9-11

Water at 20 degrees C is flowing through the system shown below. Determine the flow rate assuming a square-edged entrance from the reservoir to the pipe. Use Moody diagram and include minor losses in your calculations.

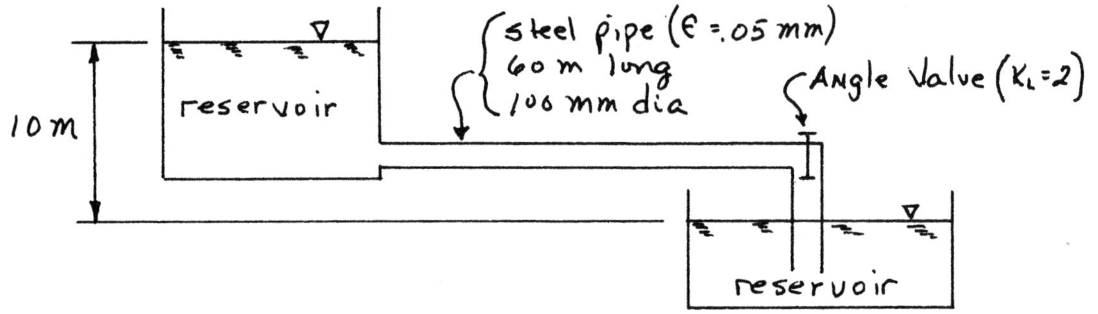

****

let point ① be the water surface at upper reservoir
let point ② be the water surface at lower reservoir
let water surface of lower reservoir be the datum line.

$$\frac{V_1^2}{2g} + z_1 + \frac{P_1}{\gamma} - h_f - h_L = \frac{V_2^2}{2g} + z_2 + \frac{P_2}{\gamma}$$

$$0 + 10\text{m} + 0 - h_f - h_L = 0 + 0 + 0 \Rightarrow h_f + h_L = 10\text{m}.$$

$$h_f = f\frac{L}{D}\frac{V^2}{2g} \qquad h_L = \Sigma K_L \frac{V^2}{2g}$$

$$K_L = \begin{cases} 0.5 \text{ (square-edged entrance)} \\ 2.0 \text{ (angle valve)} \\ 1.0 \text{ (pipe to reservoir)} \end{cases} \Rightarrow \Sigma K_L = 3.5$$

$$10\text{m} = f\frac{L}{D}\frac{V^2}{2g} + 3.5\frac{V^2}{2g} = \frac{V^2}{2g}\left[f\,\frac{60\text{m}}{0.1\text{m}} + 3.5\right]$$

$$V^2 = \frac{(10\text{m})(2\times 9.8\text{ m/sec}^2)}{600f + 3.5} \Rightarrow V = \sqrt{\frac{196\text{ m}^2/\text{sec}^2}{600f + 3.5}} \quad \text{Eq ①}$$

$$N_R = \frac{V\cdot D}{\nu} = \frac{V(0.1\text{ m})}{1.003\times 10^{-6}\text{ m}^2/\text{sec}} \Rightarrow N_R = 10^5\,V \quad \text{Eq ②}$$

TRIAL ① assume $f = .02$

$$\text{Eq ① } V = \sqrt{\frac{196}{600(.02) + 3.5}} = 3.56\text{ m/sec}$$

Eq ② $N_R = 10^5(3.56) = 3.56 \times 10^5$

$$\frac{\epsilon}{D} = \frac{.05\,mm}{100\,mm} = .0005$$

$\therefore f$ (from Moody diagram) $= .018$ N.G. !!

TRIAL ② assume $f = .018$

Eq ① $V = \sqrt{\dfrac{196}{600(.018) + 3.5}} = 3.70\,m/sec$

Eq ② $N_R = 3.7 \times 10^5$

$$\frac{\epsilon}{D} = .0005$$

$\therefore f$ (from Moody diagram) $= .018$ O.K. !!

$\therefore V = 3.70\,m/sec$

$Q = VA = 3.70\,\frac{m}{sec} \times \frac{\pi}{4}(0.1m)^2 \Rightarrow Q = 0.029\,m^3/sec$

9-12

A single pipe inverted siphon transports water at 50°F beneath the Bushkill Creek at Easton, PA as shown below. The piping consists of 200 ft. of 6-inch asphalted cast iron, including three flanged 90° elbows and an angle valve. Determine the pressure (in ft. H_2O) at point 2 when the flow rate through the siphon is 1.5 cfs.

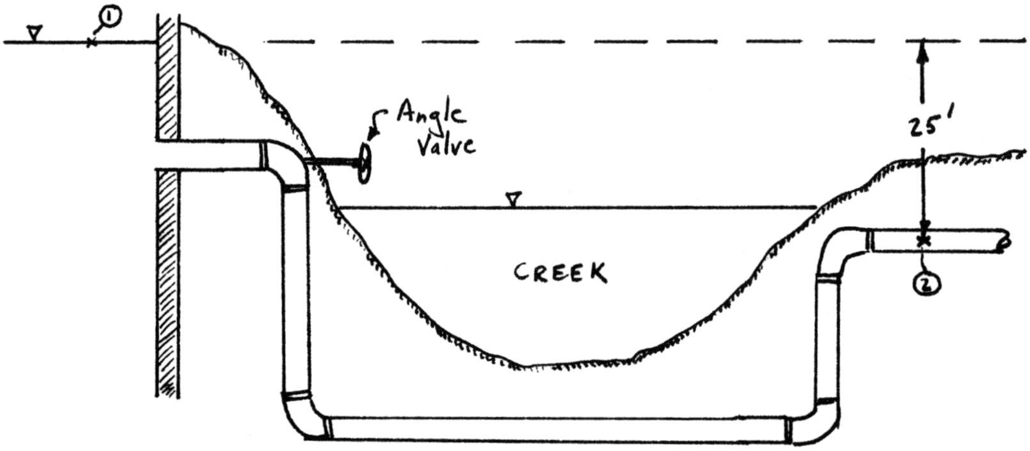

****

Bernoulli balance between points ① & ② :

$$\frac{p_1}{\gamma}^{0} + z_1 + \frac{V_1^2}{2g}^{0} = \frac{p_2}{\gamma} + z_2^{0} + \frac{V_2^2}{2g} + h_L$$

$$\frac{p_2}{\gamma} = z_1 - \frac{V_2^2}{2g} - h_L \quad \boxed{I}$$

$$V_2 = \frac{Q}{A_2} = \frac{1.5}{\pi(.25)^2} = 7.64 \text{ fps} \; ; \; \frac{V_2^2}{2g} = 0.90$$

$$\nu = 1.41 \times 10^{-5} \text{ ft}^2/\text{s from tables}$$

$$\mathbb{R}_e = \frac{VD}{\nu} = \frac{7.64(0.5)}{1.41 \times 10^{-5}} = 2.71 \times 10^5 \; ; \; \frac{k_s}{D} = .0008$$

From Moody Diagram: $f = .020$

Minor Losses:

Entrance	$K = 0.5$
Angle Valve	$= 2.0$
3 - 90° Elbows (3 × 0.3)	$= 0.9$
	$K_T = 3.4$

$$h_L = \left[f\frac{L}{D} + K_T\right]\frac{V^2}{2g}$$

$$= \left[.020\left(\frac{200}{0.5}\right) + 3.4\right]\frac{(7.64)^2}{64.4} = 10.3'$$

substitute back into $\boxed{I}$:

$$\frac{P_2}{\gamma} = 25' - 0.9' - 10.3' = \underline{\underline{13.8 \text{ ft. } H_2O}}$$

9-13

Water at 40° C is flowing. All pipe is 100 mm ID and the total length of the pipe is 100 m. Velocity is 2.0 m/s and D/ϵ = 500. The globe valve is fully open. Calculate the total head loss.

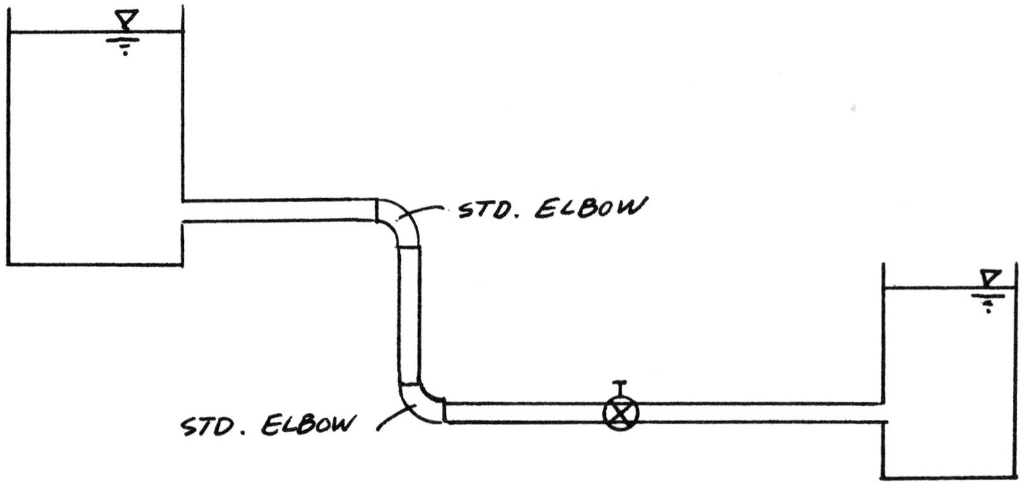

****

$$N_R = \frac{VD}{\nu} = \frac{2.0(.1)}{6.56 \times 10^{-7}} = 305{,}000 \; ; \; f = .0245$$

(MOODY'S DIAGRAM)

ENTRANCE: $h_1 = C\frac{V^2}{2g} = \frac{(.50)(2)^2}{2 \times 9.81} = .10 \text{ m}$

2 ELBOWS: $h_2 = 2f\left(\frac{L}{D}\right)\frac{V^2}{2g} = \frac{2(.0245)(30)(2)^2}{2 \times 9.81} = .30 \text{ m}$

Pipe: $h_3 = \frac{fLV^2}{D2g} = \frac{(.0245)(100)(2)^2}{(.1)(2 \times 9.81)} = 4.99 \text{ m}$

GLOBE VALVE: $h_4 = f\frac{L}{D}\frac{V^2}{2g} = \frac{(.0245)(340)(2)^2}{2(9.81)} = 1.70 \text{ M}$

EXIT: $h_5 = \frac{V^2}{2g} = \frac{(2)^2}{2(9.81)} = .20 \text{ m}$

TOTAL = 7.29 M ANS

PUMPS AND TURBINES IN REAL SYSTEMS

9-14

Draw the energy profile of the system in the figure for a flow of Q = 100 GPM of water (62.4 lb/ft^3) and f = 0.027. What horsepower is the pump releasing to the system?

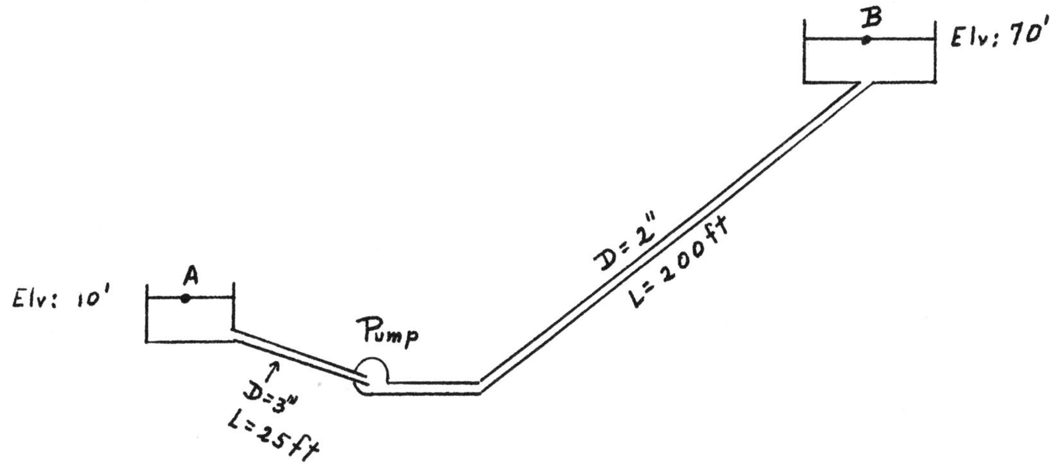

****

At A Energy 10 ft

At Pump

$$h_L = f \frac{L}{D} \frac{v^2}{2g}$$

$$= 0.027 \frac{25}{3/12} \times \frac{v^2}{2 \times 32.2}$$

$$Q = \frac{100\,gal}{min} \times \frac{1\,min}{60\,s} \times \frac{1\,ft^3}{7.48\,gal} = 0.223\,ft^3/s$$

$$h_L = 0.86\,ft$$

$$vel = \frac{Q}{A} = \frac{0.223}{\pi (1/4)^2/4} = 4.54\,ft/s$$

Energy $10\,ft - 0.86\,ft = 9.14\,ft$

At B

losses between Pump and B

$$h_L = f \frac{L}{D} \frac{v^2}{2g} \qquad v = \frac{Q}{A} = \frac{0.223}{\pi (1/6)^2/4} = 10.22\ ft/s$$

$$h_L = 0.027 \frac{200\ ft}{(1/6)\ ft} \frac{(10.22\ ft/s)^2}{2 \times 32.2\ ft/s^2}$$

$$= 52.54\ ft$$

Energy Profile:

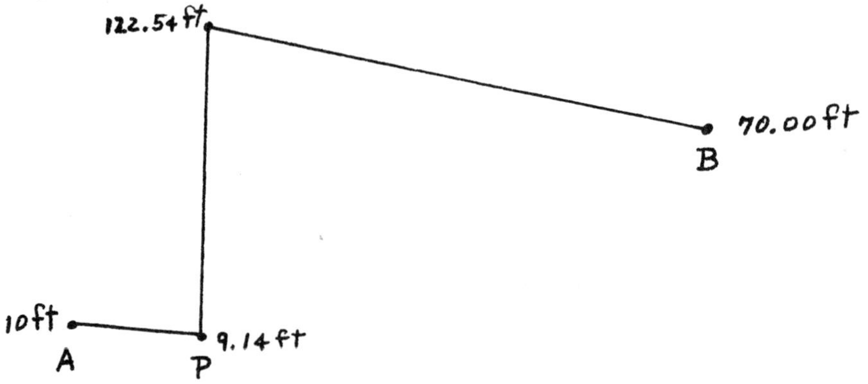

$$P = Qh\gamma = 0.223 \frac{ft^3}{s} \times h \times 62.4\ lb/ft^3 \times \frac{1\ hp}{550\ \frac{lb\ ft}{s}}$$

$$h = z_b - z_a + h_L$$

= 70 ft − 10 ft + ~~(9.14 ft)~~ + 0.86 ft + 52.54 ft

$$= 113.40\ ft$$

P = 2.87 horsepower

9-15

Water is required at a rate of 10 ft^3/s in a mountain house which has an elevation of 30 ft relative to a lake nearby. Pump efficiency is 0.75. The pipes used in the system are made of steel. Other details are given in the figure below. Assume a temperature of 60°F. Determine the power required by the pump.

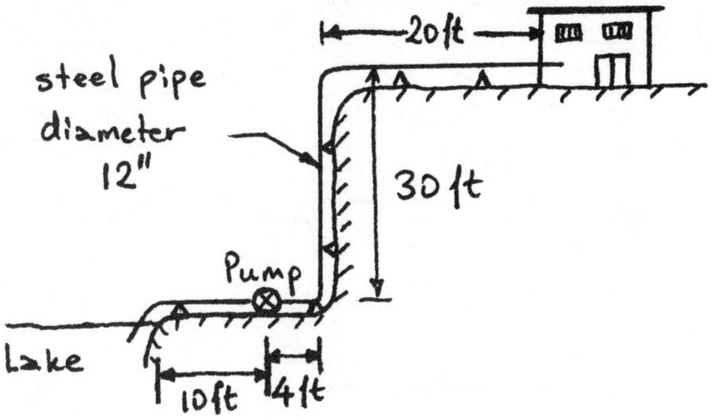

****

Pump efficiency : $\eta_p = 0.75$

Energy equation : $\frac{P_1}{\gamma} + \frac{V_1^2}{2g_c} + z_1 + h_p = \frac{P_2}{\gamma} + \frac{V_2^2}{2g_c} + z_2 + \sum h_L$

$$\frac{P_a}{\gamma} + 0 + 0 + h_p = \frac{P_a}{\gamma} + 30 + \frac{V_2^2}{2g}\left(0.5 + K_b + K_b + \frac{f.L}{D} + 1\right)$$

Sharp corners, without vanes : $K_b = 1.1$

For 60°F , $\nu = 1.22 \times 10^{-5}$ ft^2/sec , $\gamma = 62.37$ lb/ft^3

Velocity $V_2 = \frac{Q}{A} = \frac{10}{\pi (0.5)^2} = 12.73$ ft/sec.

Reynolds number $Re = V_2 . D / \nu = 12.73 \times 1 / (1.22 \times 10^{-5})$

$$Re = 1.04 \times 10^6$$

Relative roughness for steel pipe is $k_s/D = 0.00014$

Therefore, using Moody's charts we can get $f = 0.014$, $\frac{f.L}{D} = 0.896$

Energy eqn. $E_p = 30 + \frac{12.73^2}{2 \times 32.174}(0.5 + 1.1 + 1.1 + 0.896 + 1) = 41.57$ ft

Power required : $\frac{\dot{m} \times E_p}{\eta_p . 550} = \frac{10 \times 62.37 \times 41.57}{0.75} . \frac{1}{550} = 62.85$ hp .

9-16

What horsepower is the pump P delivering to the system in the Figure? The liquid being pumped is water (62.4 lb/ft^3) The flow is 400 GPM. Use the following equivalent lenghts: 2 elbows = 30 each; 2 valves = 135 each Entrance loss = C_l = 0.5 exit loss = C_l = 1.0

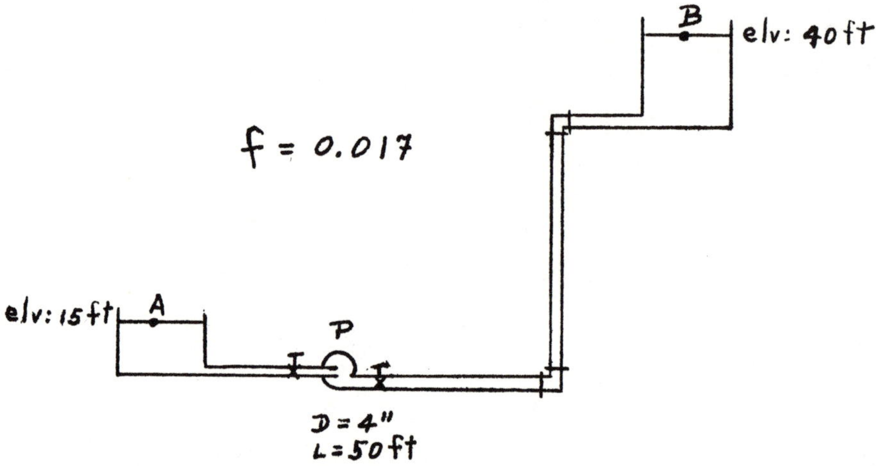

****

$$h_a = (z_b - z_a) + h_L = 40\,ft - 15\,ft + h_L$$

$$= 25\,ft + h_L$$

$$h_L = h_{pipe} + h_{Elbow} + h_{valvs} + h_{Entrc} + h_{EXIT}$$

$$= f\frac{L}{D}\frac{v^2}{2g} + 2f\,30\frac{v^2}{2g} + 2f\,135\frac{v^2}{2g} + 0.5\frac{v^2}{2g} + 1.0\frac{v^2}{2g}$$

$$v = \frac{Q}{A} = \frac{400\,gal}{min} \times \frac{1}{\pi(4/12\,ft)^2/4} \times \frac{1\,ft^3}{7.48\,gal} \times \frac{1\,min}{60\,s} = 10.2\,ft/s$$

$$h_L = \frac{(10.2)^2}{2 \times 32.2}\left(0.017\left(\frac{50}{1/3} + 60 + 270\right) + 1.5\right) = 15.61\,ft$$

$$h_a = 25\,ft + 15.61\,ft = 40.61\,ft$$

$$P = Qh\gamma = 400\frac{gal}{min} \times 40.61\,ft \times 62.4\frac{lb}{ft^3} \times \frac{1\,hp}{550\,\frac{lbft}{s}}$$

$$= 4.1 \text{ horsepower}$$

9-17

Water is being pumped from reservoir *A* to reservoir *B*. The pump delivers 70 HP, the velocity in the 12-inch diameter pipe is 10 ft/sec, the friction factor for the 12-inch pipe is 0.02. There is no friction loss in the 24-inch pipe. Compute the elevation of point B.

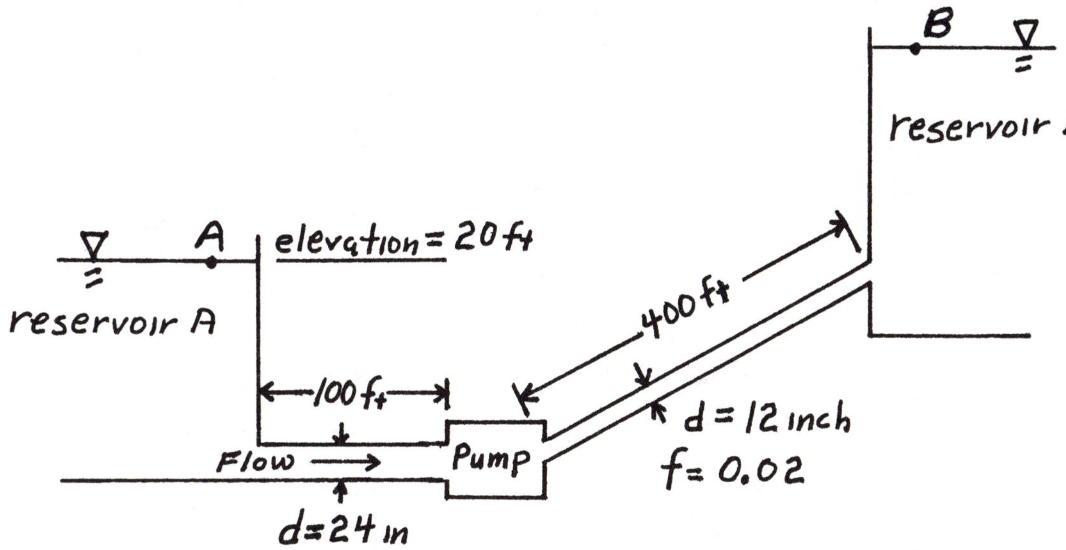

****

Apply the energy equation between points A and B.

$$\frac{P_A}{\gamma} + Z_A + \frac{V_A^2}{2g} + h_A - h_L = \frac{P_B}{\gamma} + Z_B + \frac{V_B^2}{2g}$$

$$0 + 20\,ft + 0 + h_A - h_L = 0 + Z_B + 0$$

$$Z_B = 20\,ft + h_A - h_L$$

$P_A = h_A \gamma Q$

$h_A = P_A / \gamma Q$

$Q = (\text{Velocity})(\text{Area})$

$Q = \left(10\,\frac{ft}{sec}\right)\left(\frac{\pi \cdot 1^2}{4}\right) ft^2$

$Q = 7.85\ ft^3/s$

$h_L = f \frac{L}{D} \frac{V^2}{2g}$

$h_{L_{12}} = (0.02)\left(\frac{400\ ft}{1\ ft}\right) \frac{(10\ ft/s)^2}{(2)(32.2\ ft/s^2)}$

$h_{L_{12}} = 12.4\ ft$

$h_{L_{24}} = (0.02)\left(\frac{100}{2}\right)\frac{(2.5)^2}{(2)(32.2)} = 0.097\ ft$

$P_A = 70\ HP \times 550\ \frac{ft\text{-}lb/s}{HP} = 38{,}500\ \frac{ft\text{-}lb}{s}$

$h_A = 38{,}500\ \frac{ft\text{-}lb}{s} \Big/ (62.4\ lb/ft^3)(7.85\ ft^3/s) \ \} \ h_A = 78.6\ ft$

$Z_B = 20 + 78.6 - 12.4 - 0.1 = \underline{\underline{86.1}}\ ft$

9-18

Five hundred gallons per minute of water are pumped from a reservoir into an elevated tank. Calculate the electric power required by the pump. Draw the energy line-hydraulic grade line for the system. Assume a pump efficiency of 70 percent.

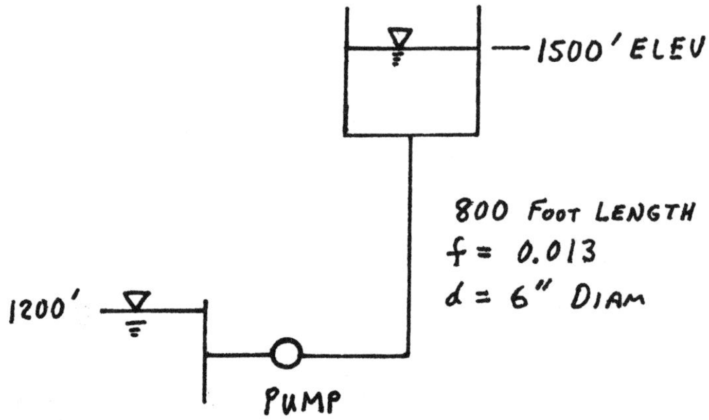

****

CALCULATE HEADLOSS IN PIPE

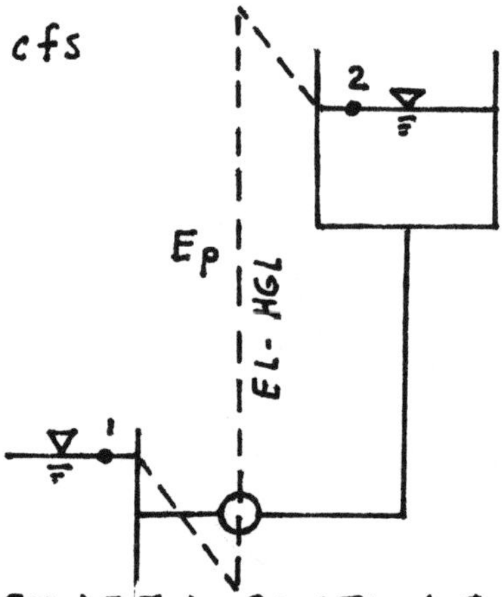

$$Q = \frac{500 \text{ gal}}{\text{min}} \; \frac{\text{min}}{60 \text{ sec}} \; \frac{ft^3}{7.48 \text{ gal}} = 1.114 \text{ cfs}$$

$$V = \frac{Q}{A} = \frac{1.114 \text{ cfs}}{\pi (.25)^2} = 5.67 \text{ fps}$$

$$h_L = f \frac{l}{d} \frac{V^2}{2g} = 0.013 \; \frac{800}{0.5} \; \frac{(5.67)^2}{(2)(32.2)}$$

$$h_L = 10.38 \text{ ft.}$$

WRITE BERNOULLI'S EQUATION BETWEEN POINTS 1 & 2

$$\frac{P_1}{\gamma} + \frac{V_1^2}{2g} + z_1 + E_P = \frac{P_2}{\gamma} + \frac{V_2^2}{2g} + z_2 + h_L$$

$$1200 + E_P = 1500 + 10.38 \qquad E_P = 310.38 \text{ ft.}$$

CALCULATE POWER REQUIRED – $P = \frac{Q \gamma E_P}{550 \, \eta_P}$

$$P = \frac{(1.114 \text{ cfs})(62.4 \text{ LB}/ft^3)(310.38 \text{ ft})}{550 \frac{\text{ft-LB}}{\text{s}}/\text{hp}} (0.746 \text{ KW}/\text{hp}) = \underline{\underline{29.3 \text{ KW}}}$$

10 CONDUIT SYSTEMS

PIPES IN SERIES

10-1

Water flows with a discharge of 2 ft^3/s from the tank on the left to the other tank on the right side.Both pipes are rough and have an equivalent sand roughness of 0.0004 ft.Assume that temperature of water is 80°F.Determine the difference in the water level of these two tanks.

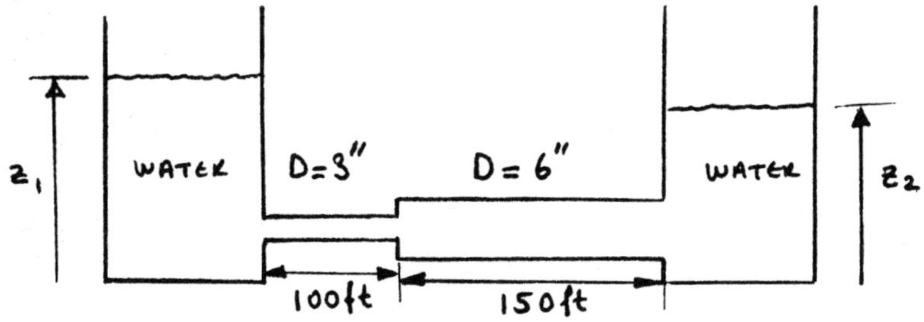

****

Energy equation can be applied between two water surfaces:

$$\frac{p_1}{\gamma} + z_1 + \frac{V_1^2}{2g} = \frac{p_2}{\gamma} + z_2 + \frac{V_2^2}{2g} + \sum h_L$$

where $p_1 = p_2 = p_{atm}$ and $V_1 = V_2 = 0$

For 100-ft pipe $h_{L_1} = \left(K_C + \frac{f_1 L_1}{D_1} + K_E\right) \frac{V_1^2}{2g}$

For 150-ft pipe $h_{L2} = \left(\frac{f_2 L_2}{D_2} + K_E\right) \frac{V_2^2}{2g}$

To determine h_{L_1}, we must determine these quantities:

$$V_1 = \frac{Q}{A_1} = \frac{2}{\pi \left(\frac{1.5}{12}\right)^2} = 40.74 \text{ ft/sec. : velocity}$$

$$k_s / D_1 = 0.0004 / 0.25 = 0.0016 \quad \text{: relative roughness}$$

$$Re_1 = \frac{V_1 D_1}{\nu} = \frac{40.74 \times 0.25}{0.93 \times 10^{-5}} = 1.095 \times 10^6 \rightarrow f_1 = 0.045$$

(at 80°F, $\nu = 0.93 \times 10^{-5}$ ft²/sec for water)

Similarly, for h_2,

$$V_2 = \frac{Q}{A_2} = \frac{2}{\pi \left(\frac{3}{12}\right)^2} = 10.19 \text{ ft/sec}, \quad k_s / D_2 = 0.0004/0.5 = 0.0008$$

$$Re_2 = \frac{V_2 D_2}{\nu} = \frac{10.19 \times 0.5}{0.93 \times 10^{-5}} = 5.48 \times 10^5 \rightarrow f_2 = 0.0194$$

Therefore,

$$h_{L_1} = \left(0.5 + \frac{0.045 \times 100}{0.25} + 0.56\right) \frac{40.74^2}{2 \times 32.17} = 491.62 \text{ ft}$$

$$h_{L2} = \left(\frac{0.0194 \times 150}{0.5} + 1.0\right) \frac{10.19^2}{2 \times 32.17} = 11.0 \text{ ft}$$

Difference in the level of the two surfaces:

$$\Delta z = z_1 - z_2 = \sum h_L = h_{L_1} + h_{L_2} = 491.62 + 11.0 = 502.62 \text{ ft}.$$

10-2

Pipes from reservoirs A, B, and C are joined at a common point. Determine the flow in each pipe.

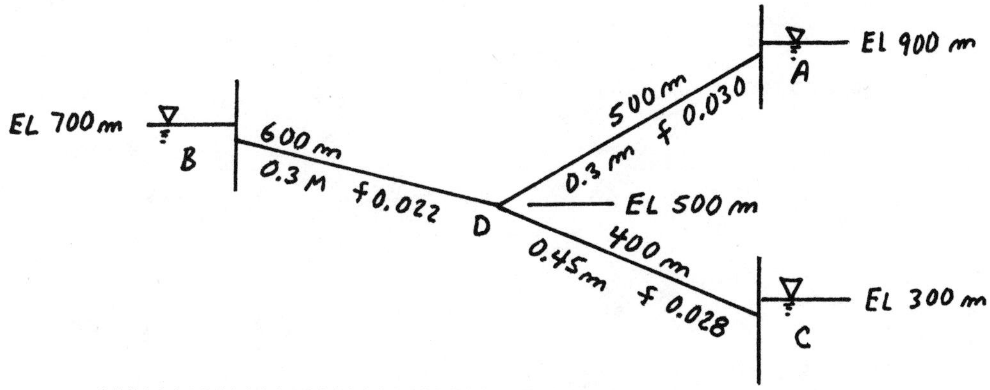

****

START ANALYSIS BY ASSUMING POINT E TO BE AT ELEVATION OF RESERVOIR B. IF $Q_{AD} > Q_{DC}$ RAISE POINT E. IF $Q_{AD} < Q_{DC}$ LOWER POINT E.

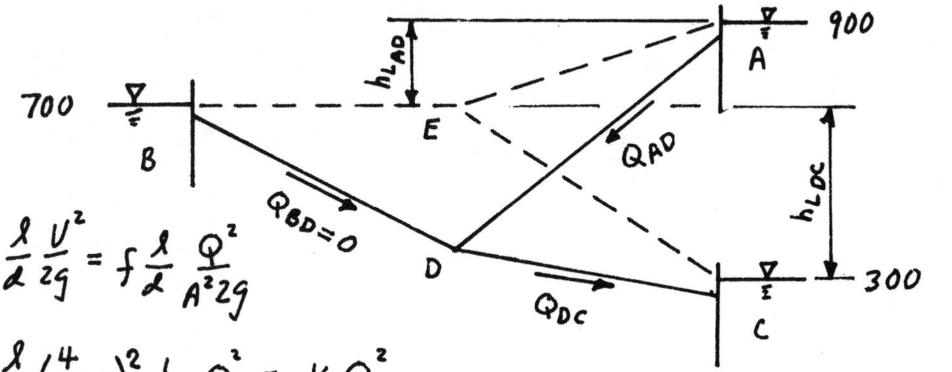

$$h_L = f\frac{l}{d}\frac{V^2}{2g} = f\frac{l}{d}\frac{Q^2}{A^2 2g}$$

$$h_L = f\frac{l}{d}\left(\frac{4}{\pi d^2}\right)^2\frac{1}{2g}Q^2 = KQ^2$$

$$K_{AD} = f\frac{l\,16}{\pi^2 d^5 2g} = 5100 \qquad K_{BD} = 449 \qquad K_{DC} = 50.2$$

$$h_{L_{AD}} = 200\text{m} = K_{AD}Q_{AD}^2 = 5100\,Q_{AD}^2 \qquad Q_{AD} = 0.198\ \text{m}^3/\text{s}$$

$$h_{L_{DC}} = 400\text{m} = K_{DC}Q_{DC}^2 = 50.2\,Q_{DC}^2 \qquad Q_{DC} = 2.82\ \text{m}^3/\text{s}$$

SINCE $Q_{AD} < Q_{DC}$ LOWER POINT E BY TRAIL AND ERROR UNTIL $Q_{AD} + Q_{BD} = Q_{DC}$

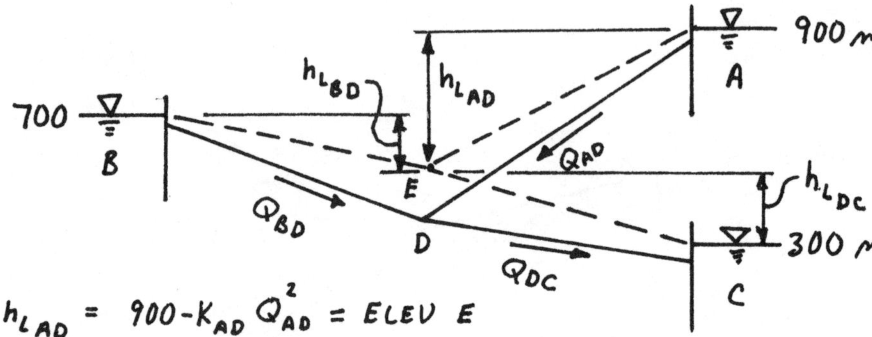

$900 - h_{L_{AD}} = 900 - K_{AD} Q_{AD}^2 = \text{ELEV } E$

T & E SOLUTION FOR

$700 - h_{L_{BD}} = 700 - K_{BD} Q_{BD}^2 = \text{ELEV } E$

$Q_{AD} + Q_{BD} = Q_{DC}$

$300 + h_{L_{DC}} = 300 + K_{DC} Q_{DC}^2 = \text{ELEV } E$

AT POINT E = ELEV. 370 m $Q_{AD} = \underline{0.322}$ $Q_{BD} = \underline{0.857 \text{ m}^3/\text{s}}$

$Q_{AD} + Q_{BD} = 0.322 + 0.857 = 1.179 \simeq Q_{DC} = \underline{1.180 \text{ m}^3/\text{s}}$

FLOWS ARE BALANCED WITHIN ACCURACY OF DATA

PIPES IN PARALLEL

10-3

One cubic meter per second of water is flow in a 0.65 meter diameter pipe. The flow is split into a parallel pipe system of 0.25 and 0.50 meter diameter pipes. Compute the flow in each pipe and the pressure drop across the parallel pipe system.

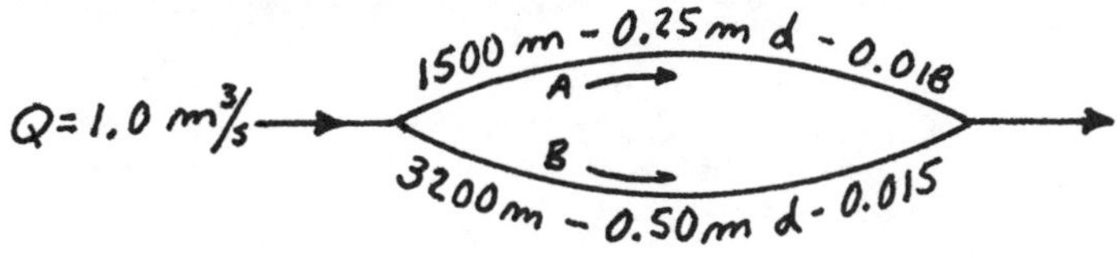

****

DIVIDE FLOWS BY SETTING

$h_{L_A} = h_{L_B} = K Q^2$

$K = f \dfrac{\ell \, 16}{\pi^2 d^5 \, 2g}$

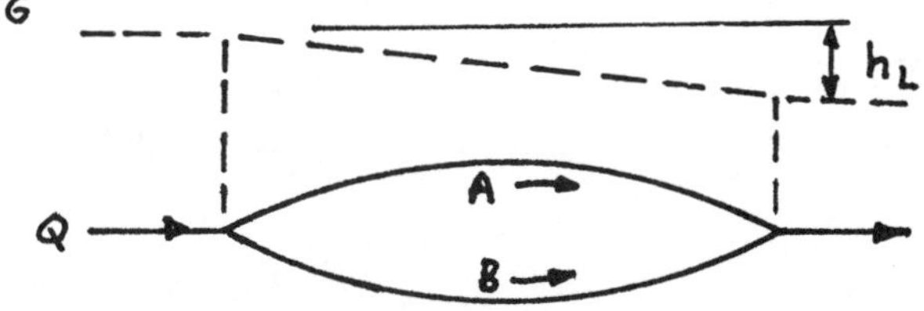

$K_A = 2284 \qquad K_B = 127 \qquad 2284\, Q_A^2 = 127\, Q_B^2 \qquad Q_B = 17.98\, Q_A$

SUBSTITUTE IN $\quad Q = Q_A + Q_B = Q_A + 17.98\, Q_A = 18.98\, Q_A = 1.0$

$Q_A = \underline{0.053\ m^3/s} \qquad Q_B = \underline{0.947\ m^3/s}$

$h_L = K_A Q_A^2 = 2284\,(0.053)^2 = \underline{6.42\ m}$

$\Delta P = \gamma h_L = (9.789\ KN/m^3)(6.42\,m) = \underline{62.8\ KPa}$

10-4

Oil (viscosity = 0.01 N•S/m^2) flows in a 0.1 m diameter pipe of 10 km length with a pressure drop of 8 MPa.

(a) Find the volume flow rate.

(b) If another pipe of equal dimensions (pipe 2) is added as shown, what is the new pressure drop if the same volume flow rate is maintained?

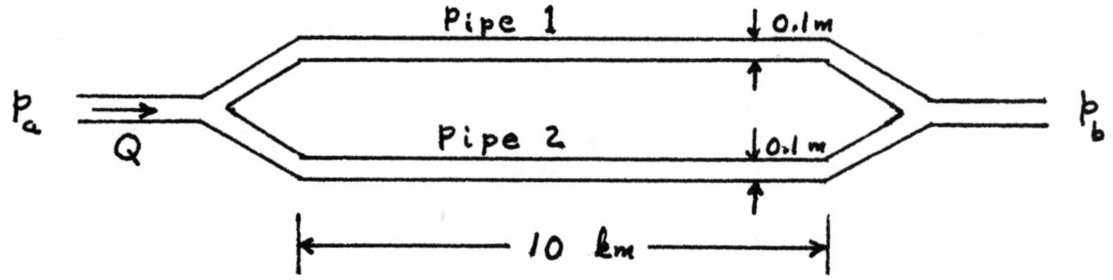

****

(a) Assuming laminar flow,

$$\text{Resistance} = R_1 = \frac{128\,\mu L}{\pi D^4} = \frac{128\,(.01)\,\frac{N\cdot s}{m^2}\,10^4\,m}{\pi(.1)^4\,m^4}$$

$$R_1 = 40.7 \times 10^6\ \frac{N\cdot s}{m^5}$$

$$Q = \frac{p_a - p_b}{R_1} = \frac{\Delta p}{R_1} = \frac{8\times10^6\ N/m^2}{40.7\times10^6\ \frac{N\cdot s}{m^5}} = 0.196\ m^3/s$$

(b) Similar to electrical resistance

$$Q = Q_1 + Q_2 = \frac{\Delta p}{R_1} + \frac{\Delta p}{R_2} = \Delta p\left(\frac{R_2+R_1}{R_1 R_2}\right) = \frac{\Delta p}{R_{Total}}$$

$$R_{Total} = \frac{R_1 R_2}{R_1+R_2} = \frac{R_1 R_1}{R_1+R_1} = \frac{1}{2} R_1 = \frac{40.7}{2}$$

$$\Delta p = R_{Total}\,Q = (.196)\,\frac{40.7}{2} = 4\ MPa$$

10-5

In the figure shown, volumetric flow rate in the main line is 10 ft^3/s. Determine the volumetric flow rate in each of the lines and the head loss from a point to the other.

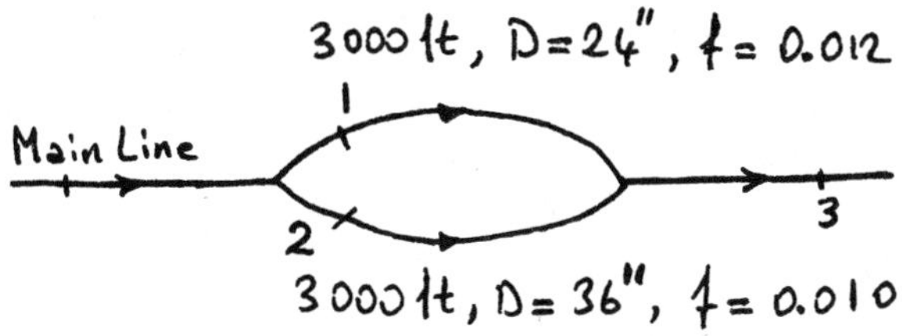

****

Continuity eqn. gives : $\rho Q_1 + \rho Q_2 = \rho Q_3 \rightarrow Q_1 + Q_2 = 10$

Head loss should be same on both sides : $h_1 = h_2$

i.e. $f_1 \cdot \frac{L_1}{D_1} \cdot \frac{V_1^2}{2g_c} = f_2 \cdot \frac{L_2}{D_2} \cdot \frac{V_2^2}{2g_c}$

Since $Q = A.V = \pi \frac{D^2}{4}.V$

$f_1 \cdot \frac{L_1}{D_1^5} \cdot Q_1^2 = f_2 \cdot \frac{L_2}{D_2^5} Q_2^2$ from which we can get

$$Q_1^2 = \frac{f_2}{f_1} \cdot \frac{L_2}{L_1} \cdot \frac{D_1^5}{D_2^5} \cdot Q_2^2 = \frac{0.010}{0.012} \times \frac{3000}{3000} \times \frac{2^5}{3^5} Q_2^2$$

$Q_1 = 0.3312\ Q_2$

Using continuity, we can obtain $Q_1 = 2.488$ ft^3/sec,
$Q_2 = 7.512$ ft^3/sec.

$$V_1 = \frac{Q_1}{A_1} = \frac{2.488}{\frac{\pi}{4} \cdot D_1^2} = \frac{4 \times 2.488}{\pi \times 2^2} = 0.792 \text{ ft/sec.}$$

$$\therefore \quad h_1 = f_1 \cdot \frac{L_1}{D_1} \cdot \frac{V_1^2}{2g_c} = 0.012 \times \frac{3000}{2} \times \frac{0.792^2}{2 \times 32.17} = 0.1755 \text{ ft.}$$

Check : $V_2 = \frac{Q_2}{A_2} = \frac{7.511}{\frac{\pi}{4} \times 3^2} = 1.0626$ ft/sec.

$$h_2 = 0.010 \times \frac{3000}{3} \times \frac{1.0626^2}{2 \times 32.17} = 0.1755 \text{ ft.}$$

10-6

Water (T=20°C) at the Easton (PA) Water Treatment Plant is pumped to the distribution system from two clearwells at the same water level elevation. The cast iron piping from the old clearwell to the pumps has an equivalent length of 200 m and a diameter of 50 cm, while the new cast iron piping is 300 m long with a 75 cm diameter. Determine the fraction of the total flow from each clearwell.

****

Parallel pipe problem

$\nu = 1 * 10^{-6}\ m^2/s$ @ 20°C from tables

Assume $Q_{OLD} = 1\ m^3/s$

$$V_{OLD} = \frac{1}{\pi(.25)^2} = 5.09\ mps$$

$$Re_{OLD} = \frac{VD}{\nu} = \frac{5.09(.5)}{10^{-6}} = 2.5 * 10^6$$

$$\frac{k_s}{D} = .00055$$

From Moody Diagram: $f = .0173$

$$h_L = f\frac{L}{D}\frac{V^2}{2g} = .0173\left(\frac{200}{0.5}\right)\frac{(5.09)^2}{19.6} = 9.15\ m$$

New piping & clearwell:

$$9.15 = f\left(\frac{300}{.75}\right)\frac{V^2}{19.6} \Rightarrow V = \sqrt{\frac{.448}{f}}$$

$$\frac{k_s}{D} = .000367$$

Iterative solution:

Assume $f = .016$, $V = 5.29$, $Re = 4 * 10^6$

New $f = .0159$, $V = 5.31$, $Re = 3.98 * 10^6$ OK

Use $V = 5.3\ mps$; $Q = VA = 5.3\left[\pi(.375)^2\right] = 2.34\ m^3/s$

$$\%\,Q_{OLD} = (1/3.34) * 100 = \underline{30\ \%}$$

$$\%\,Q_{NEW} = (2.34/3.34) * 100 = \underline{70\ \%}$$

10-7

Crude oil is flowing. Temp = 60° C, SG = 0.93. Pressure drop from A to B is 20.0 m. Length in each branch is 60 m. Use f = 0.025 as the first estimate for both pipes. Consider only pipe friction losses. All pipe is steel, schedule 40. Calculate Q in each branch.

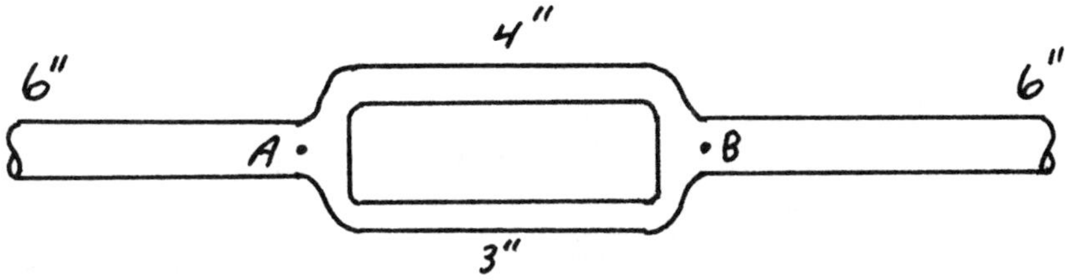

****

3" pipe: ID = .0779 m, $A = 4.768 \times 10^{-3}\ m^2$
4" pipe: ID = .1023 m, $A = 8.213 \times 10^{-3}\ m^2$
6" pipe: ID = .1541 m, $A = 1.864 \times 10^{-2}\ m^2$
} From pipe tables

DARCY'S EQN: $h_L = \dfrac{fLV^2}{D2g} = 20m$; $V = \left(\dfrac{20\,(D2g)}{fL}\right)^{\frac{1}{2}}$

4" Pipe: $V = \left(\dfrac{20\,(.1023)(2 \times 9.81)}{.025\,(60)}\right)^{\frac{1}{2}} = 5.17\ m/s$

Check Reynold's No: $N_R = \dfrac{VD\rho}{\mu} = \dfrac{5.17(.1023)(.93 \times 1000)}{1.8 \times 10^{-2}} = 27{,}000$ (μ ← GRAPH)

$\dfrac{D}{\epsilon} = \dfrac{.1023}{4.6 \times 10^{-5}} = 2200$ (ϵ ← TABLE)

f = .025 OK (From MOODY'S DIAGRAM)

$Q_{4"} = AV = (8.213 \times 10^{-3}\ m^2)(5.17\ m/s) =$ **$.0425\ m^3/s$** ANS

3" PIPE: $V = \left(\dfrac{20(.0779)(2 \times 9.81)}{(.025)(60)}\right)^{\frac{1}{2}} = 4.51\ m/s$

$N_R = \dfrac{4.51\,(.0779)(930)}{1.8 \times 10^{-2}} = 18{,}150$
$D/\epsilon = \dfrac{.0779}{4.6 \times 10^{-5}} = 1693$
} f = .026 Recalculate V = 4.43 m/s

$Q_{3"} = AV = (4.768 \times 10^{-3})(4.43) =$ **$.0211\ m^3/s$** ANS

10-8

The parallel pipe system shown is designed so that 2 cfs of water always flows through the 6 inch diameter branch with a total head loss of 3.609 feet. What diameter pipe (I.D.) is required for the bypass if minimum flow through the system is 3 cfs? Neglect losses in the tees and elbows and assume a smooth pipe.

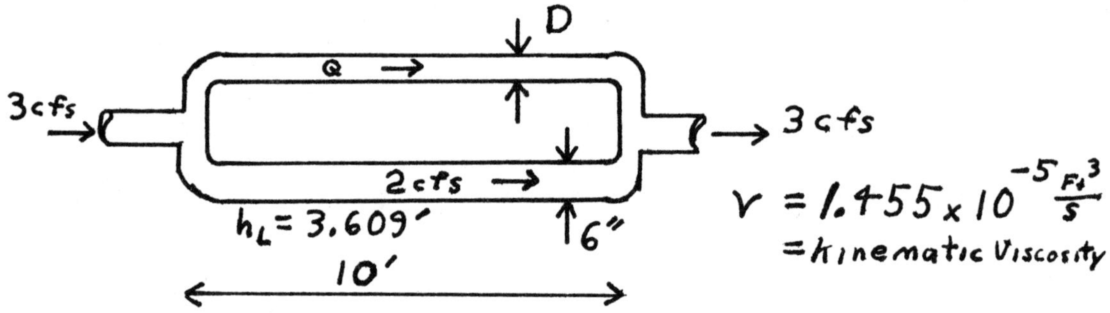

****

FROM CONTINUITY EQUATION

$$3\,cfs = 2\,cfs + Q$$
$$Q = 1\,cfs$$

FIND THE UNKNOWN VELOCITY IN TERMS OF D

$$V = \frac{Q}{A} = \frac{Q}{\frac{\pi}{4}D^2} = \frac{1\,cfs}{\frac{\pi}{4}D^2} = \frac{1.273}{D^2}\ \frac{Ft}{s}$$

KNOW: HEADLOSSES IN EACH BRANCH MUST BE EQUAL

So: h_L in the 6" branch = h_L in the unknown

$$3.609' = h_{L\ UNK.}$$

WRITE DARCY'S EQUATION FOR THE UNKNOWN PIPE

$$h_{L\ UNK.} = f \times \frac{L}{D} \times \frac{V^2}{2g}$$
$$= f \times \frac{10}{D} \times \left(\frac{1.273}{D^2}\right)^2 \times \frac{1}{2g}$$
$$= \frac{f}{D^5}(.252)$$

NOW: $$3.609 = \frac{f}{D^5}(.252)$$

$$D^5 = f(0.070)$$

$$D = \left[f \cdot 0.070\right]^{1/5}$$

SOLVE BY ITERATION WITH MOODY'S DIAGRAM

ASSUME $f = 0.02$, THEN $D = (.02 \times 0.070)^{1/5} = 0.269'$

Now Reynolds Number $(N_R) = \frac{VD}{\nu}$

$$N_R = \frac{\frac{1.273}{.269^2} \times .269}{1.455 \times 10^{-5}} = 3.25 \times 10^5$$

For Smooth Pipe MOODY'S DIAGRAM SAYS $f = 0.014$

ASSUME $f = 0.014$, THEN $D = (.014 \times 0.070)^{1/5} = 0.25'$

$$N_R = \frac{\frac{1.273}{.25^2} \times .25}{1.455 \times 10^{-5}} = 3.5 \times 10^5$$

For Smooth Pipe MOODY'S DIAGRAM SAYS $f = 0.014$ ✓

DIAMETER OF UNKNOWN PIPE = 0.25' = 3"

10-9

The piping system shown below carries water and lies in a horizontal plane. The velocity in pipe B is 7 ft/sec. Determine the flow rate in each pipe.

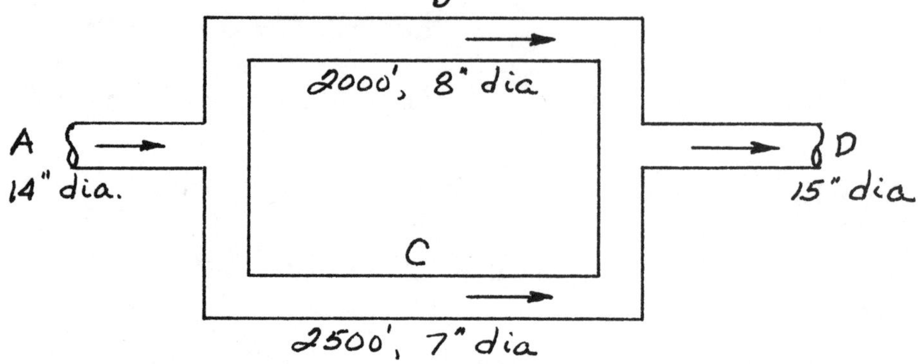

****

$$Q_A = Q_B + Q_C = Q_D$$

$$Q_B = V_B A_B = 7\,\frac{ft}{sec} \cdot \frac{\pi}{4}\left(\frac{8}{12}\,ft\right)^2 = \underline{\underline{2.44\ cfs = Q_B}}$$

$$(h_L)_B = (h_L)_C \qquad h_L = \frac{16\,fl}{2\pi^2 g d^5}\,Q^2 = KQ^2$$

$$(h_L)_B = K_B Q_B^2 \qquad (h_L)_C = K_C Q_C^2 \quad \Rightarrow K_B Q_B^2 = K_C Q_C^2$$

$$K \propto \frac{L}{D^5} \quad \Rightarrow \quad \frac{K_B D_B^5}{L_B} = \text{CONST.} = \frac{K_C D_C^5}{L_C}$$

$$\therefore K_B = K_C \frac{L_B}{L_C}\left(\frac{D_C}{D_B}\right)^5 = K_C \frac{2000\ ft}{2500\ ft}\left(\frac{7\ in}{8\ in}\right)^5 = 0.41\,K_C$$

$$K_B Q_B^2 = K_C Q_C^2$$

$$(.41 K_C)(2.44\ cfs)^2 = K_C Q_C^2 \quad \Rightarrow \quad \underline{\underline{Q_C = 1.56\ cfs}}$$

$$Q_A = Q_D = Q_B + Q_C = 2.44 + 1.56 \Rightarrow \underline{\underline{Q_A = Q_D = 4.00\ cfs}}$$

NON-CIRCULAR CONDUITS

10-10

```
Water is flowing in a rectangular culvert six feet wide and 4 feet high.
The water depth is 3.5 feet and the velocity is 8 feet per second.
Calculate the minimum head loss in 500 feet of concrete culvert
```

****

CALCULATE HYDRAULIC RADIUS – $R_h = \frac{A}{P} = \frac{6 \times 3.5}{3.5+6+3.5} = 1.615 \text{ ft.}$

SUBSTITUTE $(4R_h)$ FOR (d) WHERE REQUIRED

DETERMINE FRICTION FACTOR – f USING e/d AND REYNOLD'S No.

$e = 0.3 \text{ mm} = 0.01182 \text{ IN.}$

$$e/d = \frac{e}{(4R_h)} = \frac{0.01182 \text{ IN.}}{(4)(1.615 \text{ ft})(12 \text{ IN/ft})} = 1.52 \times 10^{-4} = 0.000152$$

$$\mathbb{R} = \frac{Vd}{\nu} = \frac{V(4R_h)}{\nu} = \frac{(8)(4)(1.615)}{(1.091 \times 10^{-5} \text{ ft}^2/\text{s})} = 4.74 \times 10^6$$

USE MOODY DIAGRAM TO FIND FRICTION FACTOR – $f = 0.013$

$$h_L = f\frac{l}{d}\frac{V^2}{2g} = f\frac{l}{(4R_h)}\frac{V^2}{2g} = (0.013)\frac{500}{(4)(1.615)}\frac{(8)^2}{(2)(32.2)} = \underline{\underline{1.00 \text{ ft}}}$$

EQUIVALENT LENGTH

10-11

A 24 inch diameter pipe branches into two pipes and then rejoins back into a 24 inch diameter pipe. One branch consists of 2000 feet of 16 inch diam. pipe connected to 1000 feet of 12 inch diam. pipe. The other branch consists of 2000 feet of 12 inch diam. pipe. Determine the equivalent length of 24 inch diameter pipe that would be required to replace the parallel pipes while maintaining equivalent operating conditions. Use a Hazen-Williams coefficient of 100 for all pipes.

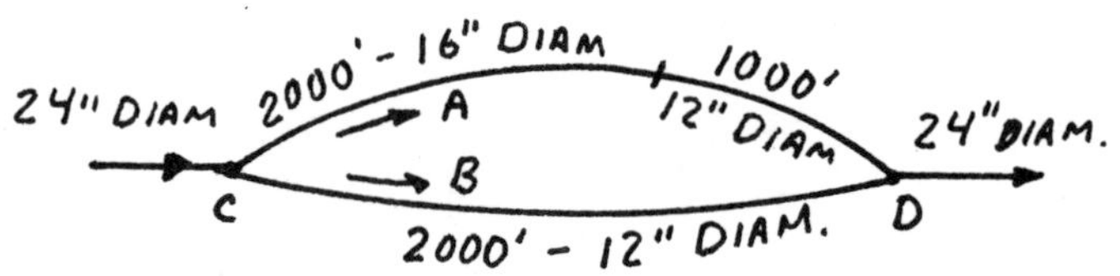

****

PROBLEMS OF THIS TYPE ARE BEST SOLVED USING A HAZEN-WILLIAMS NOMOGRAPH BUT THE SOLUTION CAN ALSO BE OBTAINED USING HAZEN-WILLIAMS EQUATIONS.

$$Q = 1.318\ A\, C_{HW}\, R_h^{0.63}\, S^{0.54} \qquad (\text{FSS UNITS})$$

SUBSTITUTING $d/4$ FOR R_h, h_L/ℓ FOR S, $\pi d^2/4$ for A AND SOLVE FOR h_L/ℓ

$$h_L(\text{ft}/1000\,\text{ft}) = 0.935\ \frac{Q^{1.85}}{d^{4.87}}$$ FOR Q IN cfs AND d IN FEET.

a) REPLACE BRANCH A BY AN EQUIVALENT PIPE OF 12 IN. DIAM.
ASSUME $Q_A = 10$ cfs. $h_{L_{16}} = 16.3$ ft/1000 ft. $h_{L_{12}} = 66.2$ ft/1000 ft.

TOTAL HEAD LOSS IN BRANCH A

$$16.3 \times \frac{2000}{1000} + 66.2 \times \frac{1000}{1000} = 98.8 \text{ ft.}$$

USING A DISCHARGE OF 10 cfs AND A HEADLOSS OF 66.2 ft/1000 FEET FOR THE 12 INCH DIAMETER PIPE, THE EQUIVALENT LENGTH OF 12 INCH PIPE IS

$$L_{12} = \frac{98.8 \times 1000}{66.2} = 1492 \text{ ft.}$$

b) ASSUME A TOTAL HEAD LOSS BETWEEN C AND D OF 50 FEET. FOR THE EQUIVALENT 12 INCH PIPE THIS YIELDS 33.5 ft/1000 ft. AND FOR THE 12 INCH PIPE THIS YIELDS 25 ft/1000 FEET. SUBSTITUTING THESE VALUES BACK INTO THE EQUATIONS YIELDS $Q_A = 6.91$ cfs AND $Q_B = 5.90$ cfs. THE TOTAL FLOW IS 12.81 cfs AT A HEAD LOSS OF 50 FEET. FOR THIS DISCHARGE, A 24 INCH PIPE WILL HAVE A HEADLOSS OF 3.58 ft/1000 FOOT OF LENGTH. THE EQUIVALENT 24 INCH PIPE TO REPLACE THE WHOLE BRANCHING SYSTEM IS:

$$\frac{50}{3.58} \times 1000 = \underline{\underline{13,966 \text{ ft}}} \text{ Long}$$

10-12

A stretch of pipe is made of 600 ft. of 10-in diameter pipe and 200 ft of 8-in diameter pipe. The attaching contraction has a C= 0.4 What is the equivalent length of the system in 8-in diam pipe? What is the relative error if the contraction is not included? Use f=0.024

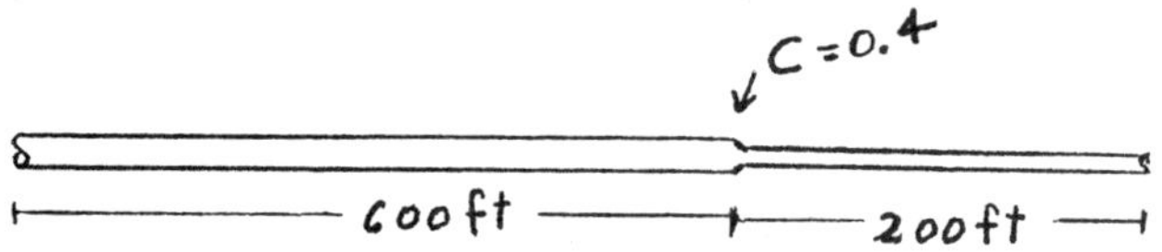

****

$$h_{10} = f \frac{L_{10}}{10''} \frac{v_{10}^2}{2g} \qquad h_8 = f \frac{L_8}{8''} \frac{v_8^2}{2g}$$

$$h_c = 0.4 \frac{v_8^2}{2g}$$

$$h_{TOTAL} = f \frac{L_{10}}{10''} \frac{v_{10}^2}{2g} + 0.4 \frac{v_8^2}{2g} + f \frac{L_8}{8''} \frac{v_8^2}{2g}$$

for any Q $\qquad A_{10} v_{10} = A_8 v_8$

$$\pi (10'')^2 v_{10} = \pi (8'')^2 v_8$$

$$10'' = 8''\left(\frac{10}{8}\right)$$

$$v_{10} = \frac{64}{100} v_8 = 0.64 v_8$$

$$h_{TOTAL} = f \frac{600\text{ft}}{8''(10/8)} \frac{(0.64 v_8)^2}{2g} + 0.4 \frac{v_8^2}{2g} + f \frac{200'}{8''} \frac{v_8^2}{2g}$$

$$= f \frac{v_8}{2g}\left(\frac{196.6'}{8''} + \frac{200'}{8''}\right) + 0.4 \frac{v_8^2}{2g}$$

$$= f \frac{v_8}{2g}\left(\frac{396.6'}{8''}\right) + 0.4 \frac{v_8^2}{2g}$$

$$= \left(\frac{396.6'}{(8''/12)} * 0.024 + 0.4\right) \frac{v_8^2}{2g} \qquad 396.6' \text{ of Eq Lgth}$$

$$= 14.27 \frac{v_8^2}{2g} \qquad \frac{0.4}{14.27} = 3\% \text{ error.}$$

PIPE NETWORKS

10-13

By the Hardy Cross iteration find the flows and velocities in each pipe for the system in the Figure. Use f=0.020

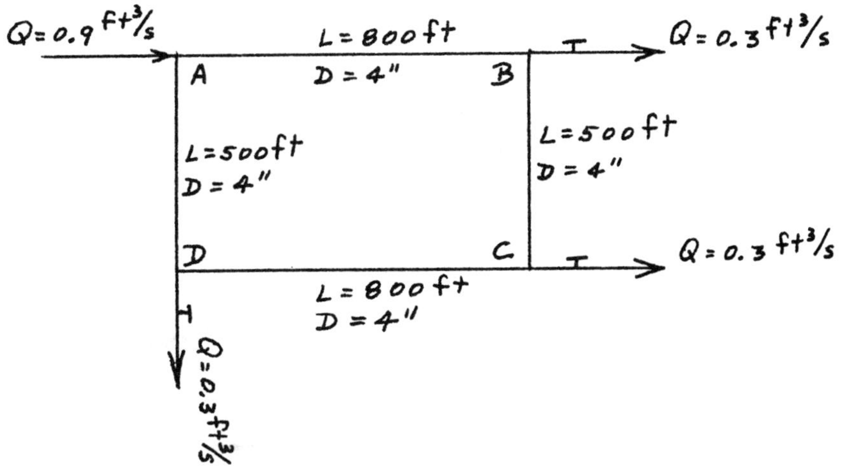

****

For losses in each pipe use $h_L = f\frac{L}{D}\frac{v^2}{2g}$

$$v = \frac{Q}{A} = \frac{Q}{\pi D^2/4} = \frac{4Q}{\pi D^2} \qquad v^2 = \frac{16Q^2}{\pi^2 D^4}$$

for A-B and C-D

$$h_L = 0.020\,\frac{800}{(1/3)^5} \times \frac{16Q^2}{\pi^2 \times 64.4} = 97.9Q^2$$

for B-C and D-A

$$h_L = 0.020\,\frac{500}{(1/3)^5} \times \frac{16Q^2}{\pi^2 \times 64.4} = 61.2Q^2$$

The easiest guess is to split the flow equally:

	KQ^2	Q	KQ^2	KQ
A − B	$97.9\,Q^2$	0.450	19.825	44.05
B − C	$61.2\,Q^2$	0.150	1.377	9.18
C − D	$97.9\,Q^2$	−0.150	−2.203	14.69
D − A	$61.2\,Q^2$	−0.450	−12.393	21.54
			+6.606	95.56

$$\Delta Q = \frac{-(6.606)}{2 \times 95.56} = -0.035$$

	new Q	KQ^2	(GPM) Q	v (ft/s)
A − B	0.415	16.861	186	4.76
B − C	0.115	0.809	52	1.32
C − D	−0.185	−3.351	~~218~~ 83	2.12
D − A	−0.485	−14.396	~~83~~ 218	5.56
		−0.077 ft		

with an error of less than 0.1 ft in pressure.

10-14

Calculate the flowrate in all the pipes in the pipe network shown.

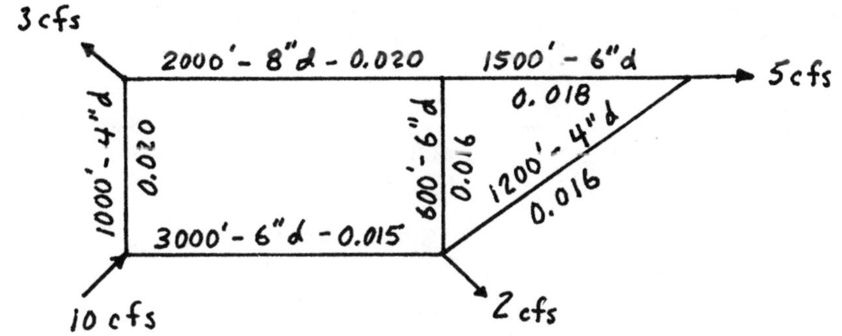

****

SET UP NETWORK.
NUMBER PIPES AND LOOPS.
ASSUME FLOWS IN PIPES.
SELECT METHOD FOR SOLVING FOR HEADLOSSES. MY CHOICE WAS $h_L = KQ^{1.85}$ BUT OTHER EQUATIONS CAN BE USED.

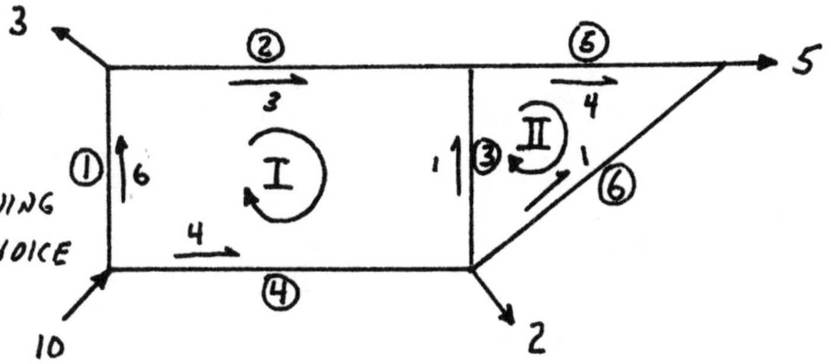

THE CORRECTION OF FLOW IS OBTAINED USING $\Delta Q = -\dfrac{\Sigma h_L}{m\Sigma(h_L/Q)}$

$K = f\dfrac{l\,16}{\pi^2 d^5\,2g}$ $\qquad m = 1.85$

TRIAL 1

LOOP NO.	PIPE NO.	DIAM (IN.)	LENGTH (FT.)	K	Q (cfs)	h_L (ft)	$\frac{h_L}{Q}$	$m\Sigma\left(\frac{h_L}{Q}\right)$	Σh_L	ΔQ
I	1	4	1000	122.3	+6	+3365.	561.			-2.25
	2	8	2000	7.65	+3	+58.4	19.5			-2.25
	3	6	600	7.73	-1	-7.73	7.73			-1.77
	4	6	3000	36.2	-4	-470.	118.	1307.	+2946	-2.25
II	3	6	600	7.73	+1	+7.73	7.73			+1.77
	5	6	1500	21.7	+4	+282.	70.5			-0.48
	6	4	1200	117.4	-1	-117.4	117.	361.	+172.	-0.48

FOR PIPE 3 WHICH IS COMMON TO BOTH LOOPS 2.25 - 0.48 = 1.77

2.25 ↑|↓ 0.48

TRIAL 2

LOOP No.	PIPE No.	K	Q (cfs)	h_L (ft)	$\frac{h_L}{Q}$	$m\Sigma(\frac{h_L}{Q})$	Σh_L	ΔQ	Q (cfs)
I	1	122.3	+3.75	+1410.	376.			−0.27	+3.48
	2	7.65	+0.75	+4.49	5.99			−0.27	+0.48
	3	7.73	−2.77	−50.9	18.4			−0.20	−2.97
	4	36.2	−6.25	−1074.	172.	1059.	+289.6	−0.27	−6.52
II	3	7.73	+2.77	+50.9	18.4			+0.20	+2.97
	5	21.7	+3.52	+223.	63.4			−0.07	+3.45
	6	117.4	−1.48	−242.	164.	455.	+31.9	−0.07	−1.55

TRIAL 3

LOOP No.	PIPE No.	K	Q (cfs)	h_L (ft)	$\frac{h_L}{Q}$	$m\Sigma(\frac{h_L}{Q})$	Σh_L	ΔQ	Q (cfs)
I	1	122.3	+3.48	+1228.	353			−0.01	+3.47
	2	7.65	+0.48	+1.97	4.10			−0.01	+0.47
	3	7.73	−2.97	−57.9	19.5			+0.01	−2.98
	4	36.2	−6.52	−1162.	178.	1026	+10.1	−0.01	−6.53
II	3	7.73	+2.97	+57.9	19.5			−0.01	+2.96
	5	21.7	+3.45	+214.	62.2			−0.02	+3.43
	6	117.4	−1.55	−264.	170.	466	+7.9	−0.02	−1.57

THE FLOWS IN THE PIPE NETWORK ARE BALANCED, BUT THE HEADLOSSES IN SEVERAL PIPES ARE EXCESSIVE. THE PIPES SHOULD BE ENLARGED AND ANALYSIS REPEATED. ALSO, PIPES WITH A LOWER FRICTION FACTOR COULD BE UTILIZED.

11 DRAG AND LIFT

VORTEX FROM CYLINDRICAL BODIES

11-1

A wire of 0.02 ft diameter is streched across the test section of a wind tunnel. Air is at 80°F inside the tunnel. Air velocity in the test section is 30 ft/s. Determine the frequency at which the vortices will shed from the cylindrical wire.

****

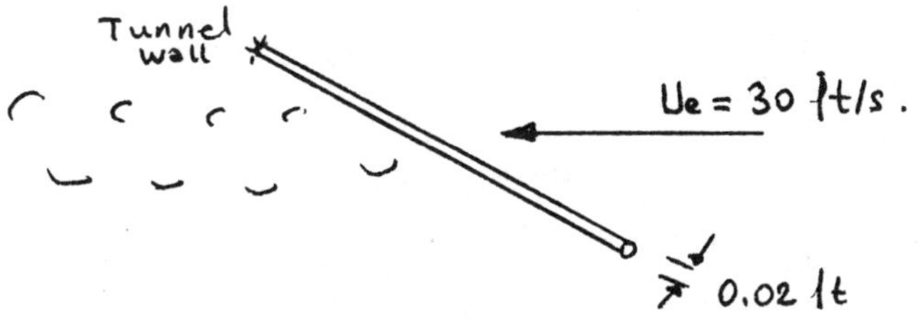

For air at 80°F, kinematic viscosity $\nu = 1.69 \times 10^{-4}\ ft^2/s$.

Reynolds number is :

$$Re_D = \frac{U_e . d}{\nu} = \frac{30 \times 0.02}{1.69 \times 10^{-4}} \cong 3550$$

For this Reynolds number, Strouhal number can be read as 0.21 from

Strouhal number ~ Reynolds number chart for cylinders (this chart can be found in almost any textbook on fluid mechanics, in the chapters on drag). Using this value, we get

$$S = \frac{n \cdot d}{U_e}$$

or $$n = \frac{S \cdot U_e}{d} = \frac{0.21 \times 30}{0.02} \cong 315 \text{ cps} .$$

FRICTION AND PRESSURE DRAG

11-2

Theoretical pressure distribution on a body is shown in the figure below. Determine the coefficient of drag for this body.

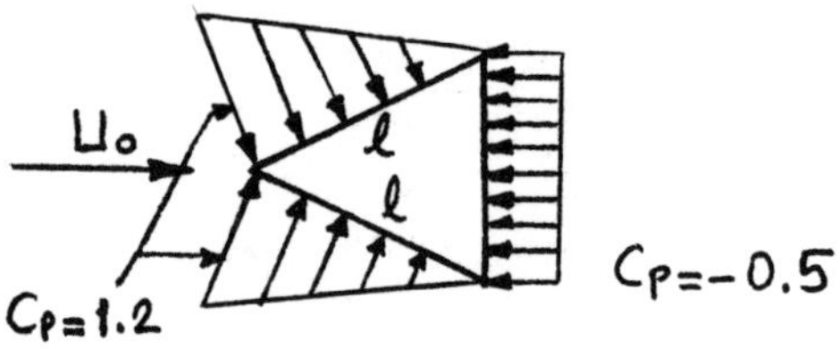

****

For the front face of this body, average C_p :

$$C_{p\,av.} = \frac{C_p \cdot l/2}{l} = \frac{C_p}{2} = \frac{1.2}{2} = 0.6$$

Drag coefficient is defined by :

$$C_D \cdot A \cdot \rho \cdot \frac{U_o^2}{2} = 2\left(C_{p\,av.} \frac{A}{2} \cdot \rho \frac{U_o^2}{2}\right)_{front} + \left(C_p \cdot A \cdot \rho \frac{U_o^2}{2}\right)_{back}$$

$$= 2\left(0.6 \frac{A}{2} \rho \frac{U_o^2}{2}\right) + \left(0.5 \cdot A \rho \frac{U_o^2}{2}\right)$$

$$= (0.6 + 0.5)\, A \cdot \rho \cdot \frac{U_o^2}{2}$$

$\therefore \quad C_D = 1.1$

11-3

A flat plate 4 ft long and 3 ft wide moves with a speed of 0.2 ft/s in water at 40°F. Determine the width of its wake at the trailing edge and the resistance force acting on this plate.

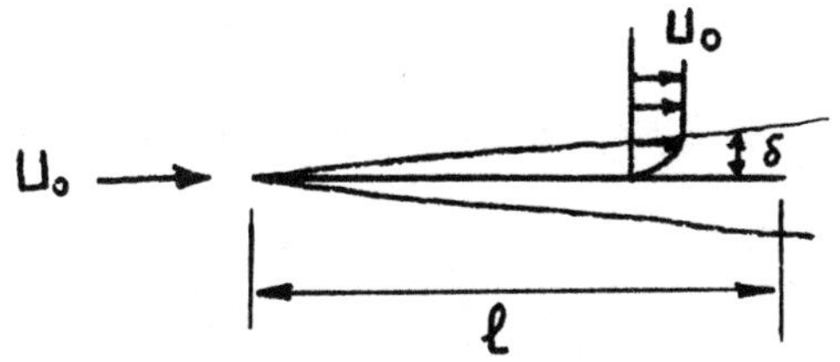

****

Reynolds number for the total length of the plate :

$$Re_L = \frac{U_o \cdot \ell}{\nu} = \frac{0.2 \times 4}{1.66 \times 10^{-5}} \cong 48200 < 5 \times 10^5$$

Laminar flow exists everywhere on this plate.

Friction coefficient for this plate : $C_L = \frac{1.33}{Re_L^{0.5}} = \frac{1.33}{219.53} = 0.00606$

Drag (resistance) force for two sides of the plate :

$$D = 2 \times (C_L . A . \rho \frac{U_o^2}{2}) = 2 \times 0.00606 \times (4 \times 3) \times 1.94 \times \frac{0.2^2}{2} = 0.00564 \text{ lbf}$$

Boundary layer thickness at the trailing edge :

$$\frac{\delta}{x} = \frac{5^*}{Re_L^{0.5}} \rightarrow \delta = 4 \times \frac{5}{219.53} = 0.0911 \text{ ft} = 1.0932 \text{ inches.}$$

Width of the wake $= 2 \times \delta = 2 \times 1.0932 = 2.1865$ inches.

*Schlichting: Boundary Layer Theory. Page 122.

DRAG IN THREE DIMENSIONS

11-4

Determine the drag force of a sphere of 6 ft diameter in air flow at 80°F. Air velocity is 50 ft/s.

****

Density and viscosity of air at 80° F (1 atm pressure) :

$$\rho = 0.00228 \ \frac{slug}{ft^3} \quad , \quad \nu = 1.69 \times 10^{-4} \ \frac{ft^2}{s}$$

Reynolds number :

$$Re = \frac{U_e . d}{\nu} = \frac{50 \times 6}{1.69 \times 10^{-4}} = 1.775 \times 10^6$$

Using $C_D \sim Re_D$ drag curve (from any textbook on fluid mechanics), for $Re \cong 1.78 \times 10^6$, we can read $C_D \cong 0.18$

Therefore drag force acting on the sphere :

$$D = C_D . \tfrac{1}{2} \rho U_e^2 . A = 0.18 \times \tfrac{1}{2} \times 0.00228 \times 50^2 \times (\pi \tfrac{6^2}{4}) = 14.5 \ lbf.$$

where $A = \pi \frac{d^2}{4}$: frontal projected area.

LIFT

11-5

A 3 ft. by 6 ft. rectangular NACA 2415 airfoil is mounted at 6° angle of attack upside down on the rear of an automobile traveling 150 ft/sec. Estimate the lift and drag forces and the horsepower consumed.

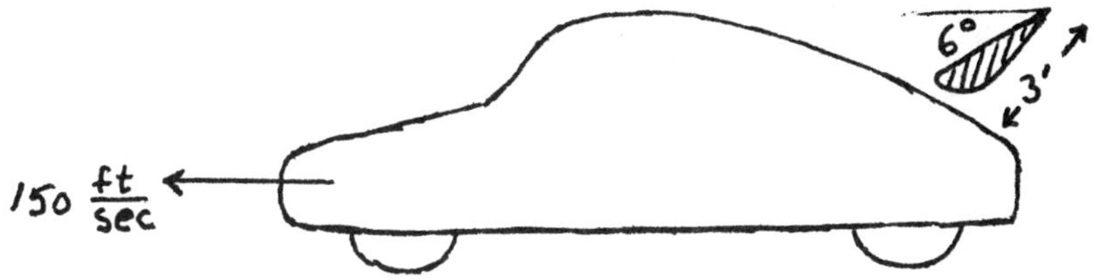

****

Assume flow over airfoil is 2-Dimensional (tips of airfoil are fenced in) and at 150 ft/sec

From the airfoil characteristic:

at $\alpha = 6°$ $\quad C_L = 0.8 \quad C_D = 0.008$

α is the angle of attack

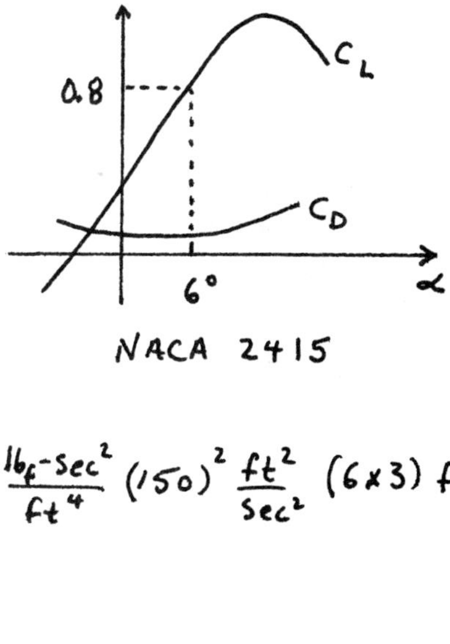

NACA 2415

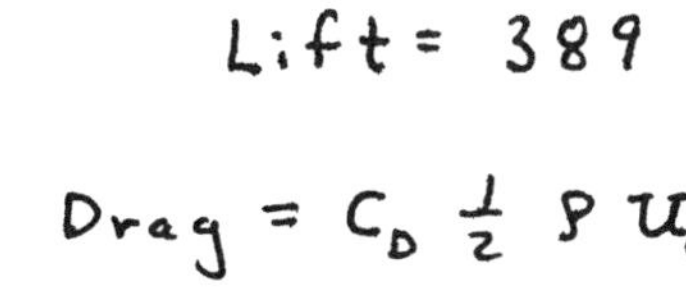

$S = bc$

$$\text{Lift} = C_L \tfrac{1}{2} \rho U_\infty^2 S = 0.8 \tfrac{1}{2} (.0024) \frac{lb_f\text{-}sec^2}{ft^4} (150)^2 \frac{ft^2}{sec^2} (6 \times 3)\, ft^2$$

Lift = 389 lb_f downward

$$\text{Drag} = C_D \tfrac{1}{2} \rho U_\infty^2 S = 0.008 \tfrac{1}{2} (.0024)(150)^2 (6 \times 3) = 39\ lb_f$$

$$\text{Horsepower Consumed by airfoil} = \frac{39\ lb_f\ 150\ ft/sec}{550\ \frac{ft\text{-}lbf}{sec\text{-}hp}} = 10.6\ hp$$

11-6

An airplane with a wing area of 8 m^2 flies horizontally at 80 m/s at sea level. Calculate the lift coefficient if the mass of the plane is 500 kg. Note that in a horizontal flight, the lift force is equal to the weight of the plane. Assume standard atmospheric conditions.

****

At standard atmospheric conditions, i.e. $P = 1$ atm. and $T = 20°C$,

$$\rho = \frac{P}{RT} = \frac{101325 \times 28.97}{8314.3 \times (20+273)} = 1.205 \ \ kg/m^3$$

∴ Lift :

$$L = C_L \cdot A \cdot \frac{1}{2} \rho U_\infty^2 = W \qquad W: \text{weight}$$

$$C_L = \frac{m g}{0.5 A \cdot \rho \cdot U_\infty^2} = \frac{500 \times 9.81}{0.5 \times 8 \times 1.205 \times 80^2} = 0.159$$

12 COMPRESSIBLE FLOW

MACH NUMBER AND CRITICAL CONDITIONS

12-1

A propeller is to be operated in air at 60°F. The diameter of this propeller is 5 ft. If the tip speed is to be kept below a value corresponding to a Mach number of 0.95, calculate the revolutions per minute allowable. If the air temperature changes to 100°F, determine whether this allowable rpm increases.

****

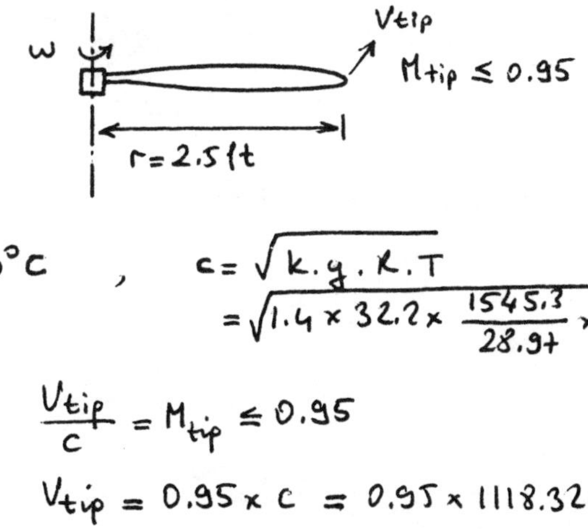

Speed of sound at 60°C , $c = \sqrt{k.g.R.T}$

$$= \sqrt{1.4 \times 32.2 \times \frac{1545.3}{28.97} \times 520} = 1118.32 \text{ ft/s}.$$

For velocity at the tip : $\frac{V_{tip}}{c} = M_{tip} \leq 0.95$

$$V_{tip} = 0.95 \times c = 0.95 \times 1118.32 = 1062.31 \text{ ft/s}.$$

$$V_{tip} = r.w = r.\frac{2\pi n}{60} \rightarrow n = \frac{V_{tip} \times 60}{2\pi \times r} = \frac{1062.31 \times 60}{2 \times 3.1416 \times 2.5} \cong 4058 \text{ rpm}$$

For 100°F , speed of sound $c = \sqrt{1.4 \times 32.2 \times \frac{1545.3}{28.97} \times 560} = 1160.43 \text{ ft/s}.$

$$V_{tip} = 0.95 \times c = 0.95 \times 1160.43 = 1102.4 \text{ ft/s}.$$

$$\therefore \quad n = \frac{1102.4 \times 60}{2 \times 3.1416 \times 2.5} \cong 4211 \text{ rpm}$$

At higher temperatures, higher rpm's are allowable.

12-2

Air flows through a nozzle with a truncated exit of 8 cm^2. The mass flow rate is 0.4 kg/s. Calculate the total pressure at the exit if the exit static temperature is 25°C and the back pressure is 120 kPa.

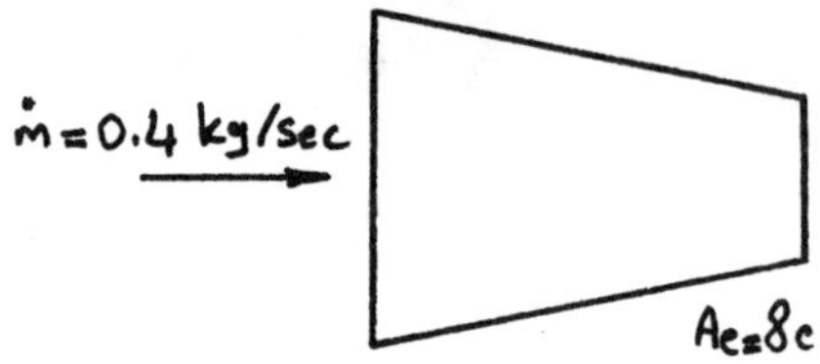

****

Speed of sound at the exit is

$$c = \sqrt{k.R.T_e} = (1.4 \times 287. \times 298)^{1/2} = 346 \text{ m/sec}$$

Velocity at the exit : $V_e = \dot{m}/\rho_e A_e = 0.5/(\rho_e \times 0.0008)$

$$\rho_e = ? \quad , \quad \rho_e = \frac{P_e}{R.T_e} = \frac{120\,000}{287 \times 298} = 1.403 \text{ kg/m}^3$$

Therefore, $V_e \cong 445.5$ m/sec

and Mach no. $M_e = \frac{V_e}{c} = \frac{445.5}{346} \cong 1.28 > 1$

This means that this nozzle is chocked.

Total temperature, using $k = 1.4$ for air

$$T_o = 1.2 \times T = 1.2 \times 298 = 357.6 \text{ °K}$$

Total pressure : $P_o = \frac{\dot{m}\sqrt{R.T_o}}{0.685\,A^*} = \frac{\dot{m}\sqrt{R.T_o}}{0.685\,A_e}$

$$= \frac{0.5 \times \sqrt{287 \times 357.6}}{0.685 \times 0.0008} = 292\,300 \text{ Pa}$$

$$= 2.92 \text{ kPa}$$

SHOCK WAVES AND WATER HAMMER

12-3

Water flows in a 5-in. diameter pipe with a velocity of 6 ft/s. The length of this pipe is 8000 ft. A valve at the end of the pipe is closed for 2 seconds. Determine the maximum force on the valve. Assume a water temperature of 60°F.

****

For water at 60°F , $\rho = 1.94$ slug/ft³

Speed of sound is

$$c = \sqrt{\frac{E_v}{\rho}} = \sqrt{\frac{320000 \times 144}{1.94}} = 4873.66 \text{ ft/s}$$

Critical time :

$$t_c = \frac{2L}{c} = \frac{2 \times 8000}{4873.66} = 3.283 \text{ s}$$

Closing time 2 s < Critical time , 3.283 s .

This will create a pressure force on the valve :

$$\Delta p = \rho . V . c = 1.94 \times 6 \times 4873.66 = 56729.4 \text{ lb}_f/\text{ft}^2$$

Therefore, maximum force on the valve :

$$F = A . \Delta p = \left(\pi \frac{d^2}{4}\right) . \Delta p = 3.1416 \times \frac{\left(\frac{5}{12}\right)^2}{4} \times 56729.4 = 7735.3 \text{ lb}_f$$

12-4

Determine the static pressure and temperature behind a shock wave occuring in a flow of air if temperature and pressure at the inlet (in front of the shock wave) are -30°C and 85 kPa. Velocity is 500 m/s in front of the wave. Determine the entropy increase through the shock wave.

****

For air $k = 1.4$, $R = 287$.

In front of the shock wave $T_1 = 273 - 30 = 243°K$

$$C_1 = \sqrt{k.R.T_1} = \sqrt{1.4 \times 287 \times 243} = 312.47 \text{ m/sec} \quad \text{speed of sound.}$$

Therefore, Mach number is $M_1 = V_1/c_1 = 500/312.47 = 1.6$ in front of the shock wave.

Mach number behind the shock wave :

$$M_2^2 = \frac{(k-1)M_1^2 + 2}{2kM_1^2 + 1 - k} = \frac{(1.4-1)\times 1.6^2 + 2}{2\times 1.4 \times 1.6^2 + 1 - 1.4} = 0.447$$

Temperature and pressure behind the shock wave:

$$T_2 = \left(\frac{1 + \frac{k-1}{2}M_1^2}{1 + \frac{k-1}{2}M_2^2}\right) T_1 = \frac{1 + \frac{1.4-1}{2}\times 1.6^2}{1 + \frac{1.4-1}{2}\times 0.447^2} \times 243 = 353.34 \text{ °K}$$

$$P_2 = \left(\frac{k M_1^2 + 1}{k M_2^2 + 1}\right) P_1 = \frac{1.4\times 1.6^2 + 1}{1.4 \times 0.447^2 + 1} \times 85 \cong 304.5 \text{ k Pa}$$

Entropy increase through the shock wave :

$$\Delta s = R \ln\left[\left(\frac{T_2}{T_1}\right)^{\frac{k}{k-1}} \cdot \frac{P_1}{P_2}\right] = 287 \times \ln\left[\left(\frac{353.34}{243}\right)^{\frac{1.4}{0.4}} \cdot \frac{85}{304.5}\right] = 9.84 \ \frac{J}{kg\,°K}$$

13 FLOW MEASUREMENTS

MEASUREMENTS OF VELOCITY AND PRESSURE

PITOT AND CURRENT METERS

13-1

What is the volumetric flow rate through the orifice meter shown below ?

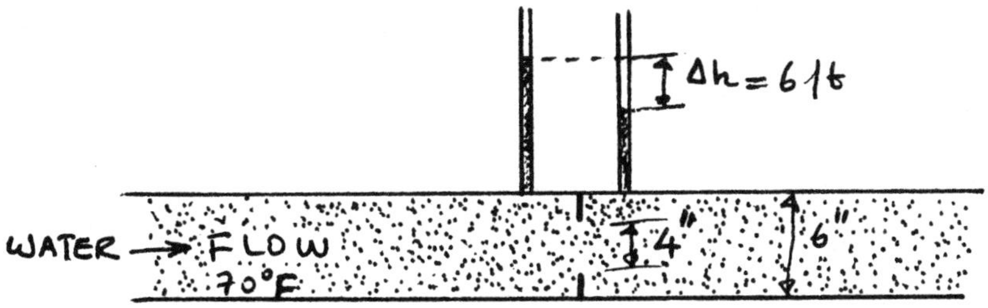

****

$$\frac{Re.d}{K} = \sqrt{2.g.\Delta h}\cdot\frac{d}{\nu}$$

$$= \sqrt{2\times 32.2\times 6}\cdot\frac{(4/12)}{1.06\times 10^{-5}} = 618\,146.55$$

$$\frac{d}{D} = \frac{4}{6} \approx 0.68$$

From flow coefficient charts , $K \approx 0.72$

$$Q = K.A_0\sqrt{2g.\Delta h} = 0.72\times\frac{\pi}{4}\times\left(\frac{4}{12}\right)^2\times\sqrt{2\times 32.2\times 6} = 1.235 \text{ ft}^3/\text{s}$$

13-2

The sketch shows the way a hydrant test is made to measure the volume flow rate of water. The hydrant is allowed to run wide open. The pressure is measured as shown with a pitot tube. The stream of water will tend to contract slightly as it flows out of the 2-1/4" ID opening in the hydrant. Assume the area of the stream of water at the pitot tube equals .9x the area of the opening.

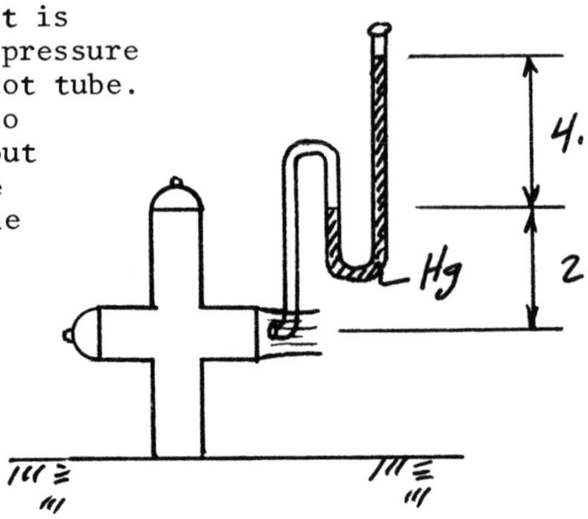

Determine Q.

****

$$A = .9\,\frac{\pi}{4}\left(\frac{2.25}{12}\right)^2 = .0249\ ft^2$$

$$\frac{P_1}{\gamma} + \frac{V_1^2}{2g} + z_1 = \frac{P_2}{\gamma} + \frac{V_2^2}{2g} + z_2 :$$

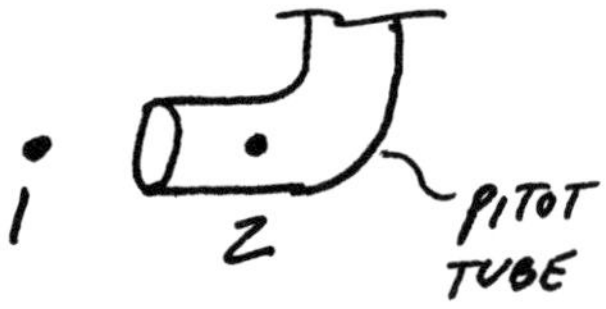

$P_1 = 0$ (Atmospheric)

$z_1 = z_2$

$V_2 = 0$ (static inside pitot tube)

$$\frac{P_2}{\gamma} = 2.3 + 4.62(13.56) = 64.9 \text{ ft. of water}$$

$$\frac{V_1^2}{2g} = 64.9\ ,\quad V_1 = \sqrt{(64.9)(2\times 32.2)} = 64.7\ \frac{ft}{s}$$

$$Q = AV = (.0249)(64.7) = 1.61\ \frac{ft^3}{sec}\quad \text{ANS}$$

13-3

A pitot-static tube and differential manometer using water at 25°C as the gage fluid will be used to measure the velocity of air at 20°C and atmospheric pressure. For a manometer deflection of 4.0 cm determine the air velocity.

****

ENERGY EQUATION:

$$\frac{P_1}{\gamma} + \frac{V_1^2}{2g} + Z_1 = \frac{P_2}{\gamma} + \frac{V_2^2}{2g} + Z_2$$

PITOT TUBE

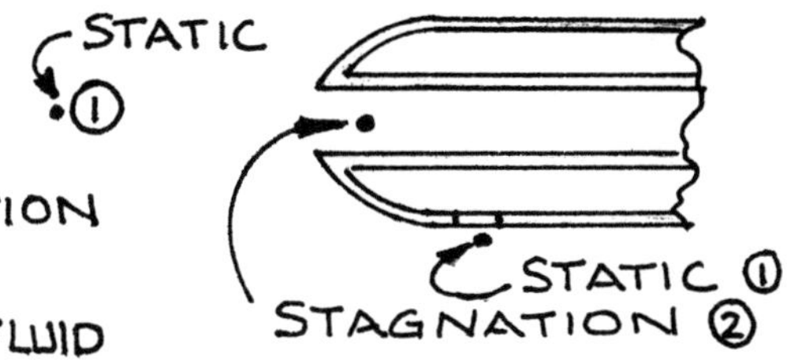

$Z_1 = Z_2$ $\quad V_2 = 0$ AT STAGNATION

$$\frac{V_1^2}{2g} = \frac{P_2 - P_1}{\gamma}$$

Mf = MEASURED FLUID

Gf = GAGE FLUID

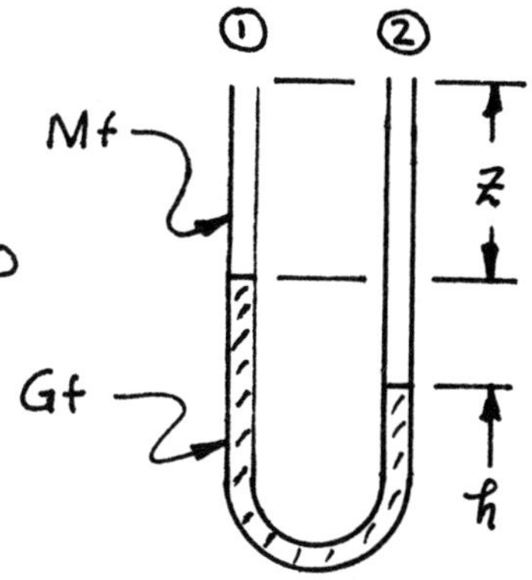

MANOMETER

FROM MANOMETER EXPRESSION:

$$P_{②} + \gamma_{Mf}(Z) + \gamma_{Mf}(h) - \gamma_{Gf}(h) - \gamma_{Mf}(Z) = P_{①}$$

$$P_{②} - P_{①} = h(\gamma_{Gf} - \gamma_{mf})$$

INSERT IN ENERGY EQUATION:

$$\frac{V_1^2}{2g} = \frac{h(\gamma_{Gf} - \gamma_{mf})}{\gamma_{mf}}$$

$$V_1 = \sqrt{2gh\left(\frac{\gamma_{Gf} - \gamma_{mf}}{\gamma_{mf}}\right)}$$

$\gamma_{Gf} = 9.78\ KN/M^3$

$\gamma_{mf} = 11.81\ N/M^3$

$$V = \sqrt{2(9.81)(.04)\left(\frac{9780 - 11.81}{11.81}\right)}$$

$\underline{V = 25.47\ M/S}$

13-4

A Pitot tube connected to a water manometer is used to measure the velocity of air flowing in a pipe. Diameter of the Pitot tube is 0.6 mm. Pressure and temperature in the pipe are 0.9 atm. and 17°C. Assume an atmospheric temperature of 20° C. For a water deflection of 1 cm on the manometer, determine the velocity.

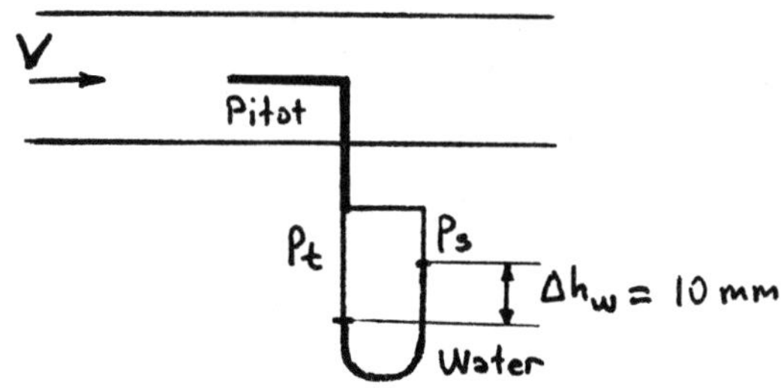

****

Total pressure $P_t = P_s + \frac{1}{2}\rho_a V^2$ $\quad$ P_s : static pressure

$$V^2 = \frac{2(P_t - P_s)}{\rho_a} = \frac{2 \cdot \Delta h_w \cdot g \cdot \Delta\rho}{\rho_a} \quad \text{where } \Delta\rho = \rho_w - \rho_a$$

$$\rho_a = \frac{P_a}{RT_a} = \frac{101325 \times 0.9}{287.06 \times 290} = 1.095 \text{ kg/m}^3$$

At 20°c $\quad \rho_w = 998 \text{ kg/m}^3$

$$V^2 = \frac{2 \times 0.010 \times (998 - 1.095) \times 9.81}{1.095} = 178.62$$

$$V = 13.36 \text{ m/sec}$$

Reynolds number $Re = \frac{V.d}{\nu} = \frac{13.36 \times 0.0006}{1.46 \times 10^{-5}} = 549.04$

From correction figures for Pitot tubes: $C_P = 1.003$

$$V = V \times C_P = 13.36 \times 1.003 = 13.40 \text{ m/sec}$$

13-5

A pitot-static tube is inserted into a flowing water stream. Compute the velocity of the water and its flowrate. What are the magnitudes of the static and stagnation pressures?

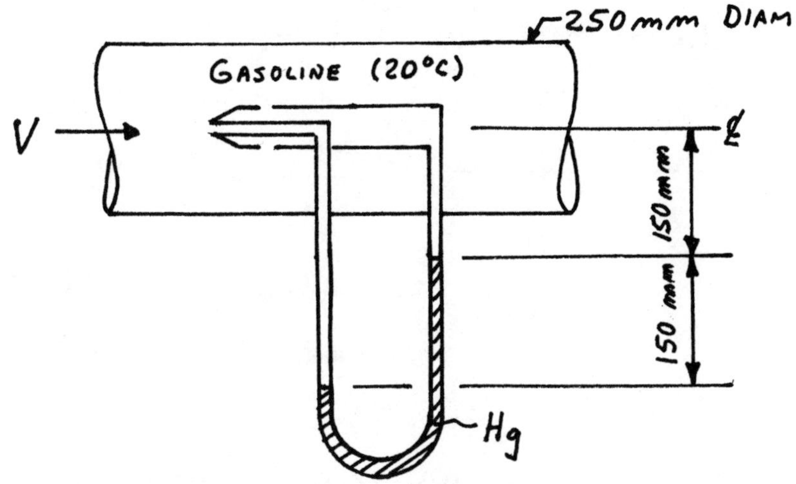

****

P_1 = STAGNATION PRESSURE

P_2 = STATIC PRESSURE

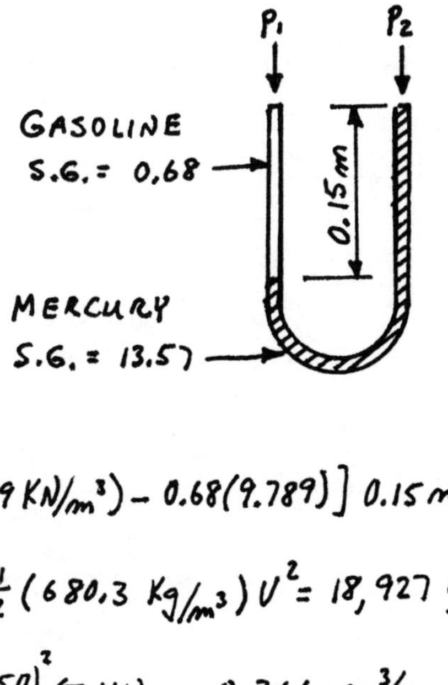

$$P_1 = P_2 + \frac{1}{2}\rho V^2$$

$$\rho = 680.3 \text{ kg/m}^3$$

$$P_1 + \gamma_{GAS} h - \gamma_{Hg} h = P_2$$

$$P_1 - P_2 = (\gamma_{Hg} - \gamma_{gas}) h = [13.57(9.789 \text{ KN/m}^3) - 0.68(9.789)]\, 0.15 \text{ m}$$

$$P_1 - P_2 = 18.927 \text{ KN/m}^2 = \frac{1}{2}\rho V^2 \qquad \frac{1}{2}(680.3 \text{ Kg/m}^3) V^2 = 18{,}927 \text{ N/m}^2$$

$$V = \underline{\underline{7.46 \text{ m/s}}} \qquad Q = AV = \frac{\pi (.250)^2}{4}(7.46) = \underline{\underline{0.366 \text{ m}^3/\text{s}}}$$

13-6

The nozzle contracts the flow down to one quarter of its orignal area. Air is the fluid, and the flow may be assumed frictionless. The upstream velocity is $v_1 = 30$ ft/sec and the upstream pressure is 10 psig. Find h for the water manometer.

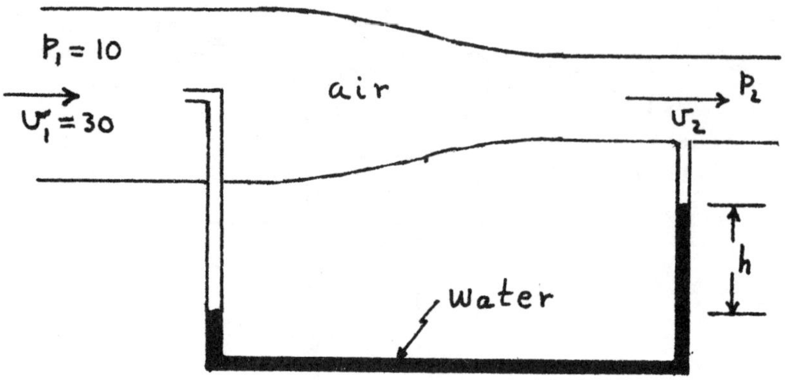

****

From Continuity, $\rho_1 A_1 V_1 = \rho_2 A_2 V_2$ but $\rho_1 = \rho_2$

$$\Rightarrow V_2 = V_1 \frac{A_1}{A_2} = 30(4) = 120 \text{ ft/sec}$$

$$P_{1\,stagnation} = P_1 + \frac{\rho V_1^2}{2} = 10 + \frac{.0024 \frac{\text{lb}_f\text{-sec}^2}{\text{ft}^4}(30)^2 \frac{\text{ft}^2}{\text{sec}^2}}{2(144)\frac{\text{in}^2}{\text{ft}^2}} = 10.0075 \frac{\text{lb}_f}{\text{in}^2}$$

For frictionless flow $P_{1\,stag} = 10.0075$ psig $= P_{2\,stag}$

$$\therefore P_2 = P_{2\,stag} - \frac{1}{2}\rho V_2^2 = 10.0075 - \frac{.0024}{2(144)}(120)^2$$

or $P_2 = 10.0075 - .12 = 9.8875$ psig

manometer measures,

$$P_{1\,stag} - P_2 = 10.0075 - 9.8875 = .12 \text{ psig} = \rho_w g h = 62.4\,h$$

$$h = \frac{.12 \frac{\text{lb}_f}{\text{in}^2}(12)^3 \frac{\text{in}^3}{\text{ft}^3}}{62.4 \frac{\text{lb}_f}{\text{ft}^3}} = 3.32 \text{ inches}$$

MEASUREMENTS OF FLOW RATE
VENTURI AND WEIRS

13-7

Water is flowing over a sharp-crested rectangular weir 1.2 m wide. The head over the weir is measured in a stilling well at an upstream section to be 250 mm. If the coefficient of discharge for the weir is 0.620, determine the rate of discharge through the weir.

****

The weir formula for discharge is:

$$Q = \frac{2}{3} b\, C_d \sqrt{2g}\, H^{3/2}$$

$$Q = \frac{2}{3}(1.2)(0.620)(\sqrt{2 \times 9.810})(0.250)^{3/2}$$

$$Q = 0.275 \text{ m}^3/\text{s}$$

13-8

Determine the flow rate of water through a triangular weir with an included angle of 90^0. The discharge coefficient for the weir is 0.6 and the head over the weir is 2.5 ft.

****

The discharge for triangular weir is given as:

$$Q = \frac{8}{15} C_d \sqrt{2g} \tan\frac{\theta}{2} (H^{5/2})$$

$$Q = \frac{8}{15}(0.6)(\sqrt{2 \times 32.174}) \times 1 \times (2.5)^{2.5}$$

or, $Q = 25.367$ cfs

13-9

A V-notch weir is used to determine the flow in a channel. Develop a relationship between flowrate and depth of flow above apex of weir. Draw a calibration curve for flow versus depth with depth on the ordinate axis. Complete for α angles of 30°, 45°, and 60°.

****

$$dQ = v\,dA = \sqrt{2gh}\,dA$$

$$dA = 2(H-h)(\tan\alpha)\,dh$$

$$dQ = \sqrt{2gh}\,(2)(H-h)(\tan\alpha)\,dh$$

$$dQ = 2\sqrt{2g}\,\tan\alpha\,(H-h)\,h^{1/2}\,dh$$

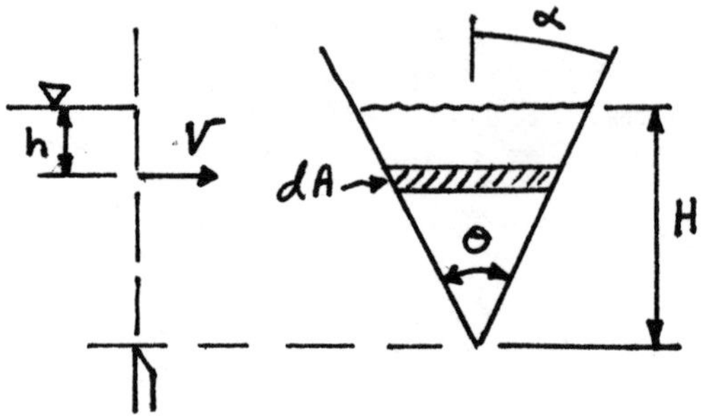

$$Q = 2\sqrt{2g}\,\tan\alpha\int_0^H (H-h)h^{1/2}\,dh = \frac{8}{15}\tan\alpha\sqrt{2g}\,H^{5/2}$$

ASSUMING $\alpha = \theta/2$, $g = 9.81\ m/s^2$ AND H IN METERS

$$Q_{60} = \frac{8}{15}\tan 30°\sqrt{2(9.81)}\,H^{5/2} = 1.364\,H^{5/2}$$

$$Q_{90} = \frac{8}{15}\tan 45°\sqrt{2(9.81)}\,H^{5/2} = 2.362\,H^{5/2}$$

$$Q_{120} = \frac{8}{15}\tan 60°\sqrt{2(9.81)}\,H^{5/2} = 4.092\,H^{5/2}$$

COMPUTE DATA POINTS:

θ	60°	90°	120°
α	30°	45°	60°
H	FLOW RATES, m^3/s		
0.2	0.024	0.042	0.073
0.4	0.138	0.239	0.414
0.6	0.380	0.659	1.141
0.8	0.781	1.352	2.342
1.0	1.364	2.362	4.092

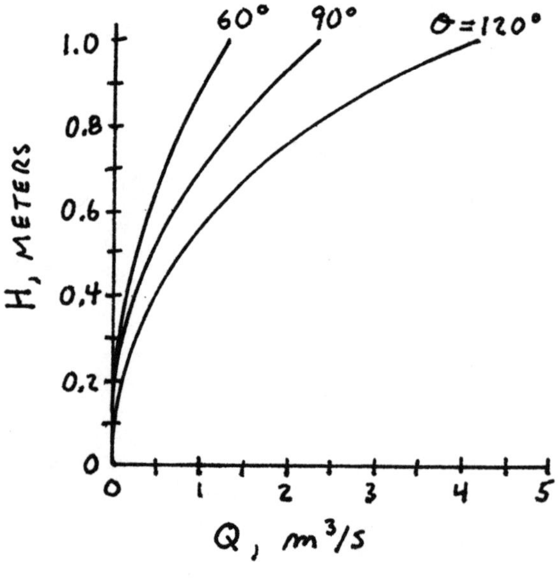

13-10

A liquid with a unit weight of 54 lb/ft^3 is flowing. What is the reading h on the manometer? Ignore any head loss.

$P_1 = 10$ PSIG
$V_1 = 12$ FT/SEC
ID = 6"

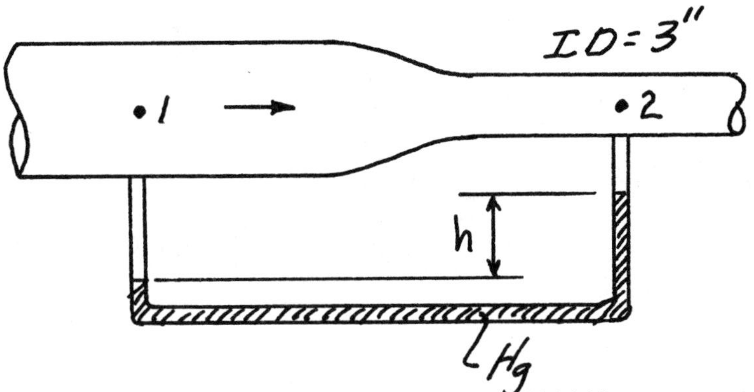

****

Manometer loop ① to ②:

$\gamma_o = 54$ pcf
$\gamma_{Hg} = 13.6\,\gamma_w$
$\gamma_w = 62.4$ pcf

$$P_1 + \gamma_o h - \gamma_w \times 13.6 h = P_2$$

$$\frac{P_2 - P_1}{\gamma_o} = h - \frac{\gamma_w \times 13.6 h}{\gamma_o} = h\left(1 - \frac{62.4 \times 13.6}{54}\right) = h(-14.71) \quad (1)$$

Bernoulli's Eqn ① to ②

$z_1 = z_2$
$V_1 = 12$ ft/sec
$V_2 = \left(\frac{6}{3}\right)^2 V_1 = 48$ ft/sec

$$\frac{P_1}{\gamma_o} + \frac{V_1^2}{2g} + z_1 = \frac{P_2}{\gamma_o} + \frac{V_2^2}{2g} + z_2$$

$$\frac{P_2 - P_1}{\gamma_o} = \frac{V_1^2}{2g} - \frac{V_2^2}{2g} = \frac{12^2 - 48^2}{2 \times 32.2} = -33.54 \text{ ft} \quad (2)$$

EQN (1) = EQN (2) $\quad h(-14.71) = -33.54$

$h = 2.28$ ft	ANS

13-11

Find the volume flow rate of water through the Venturi Meter shown. Give your answer in ft^3/sec.

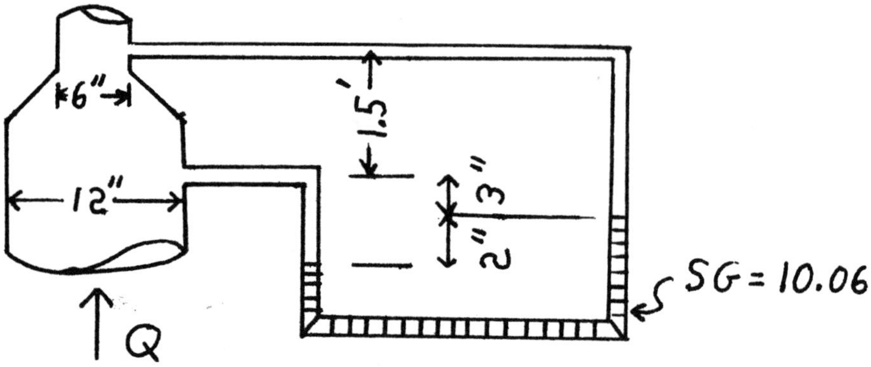

****

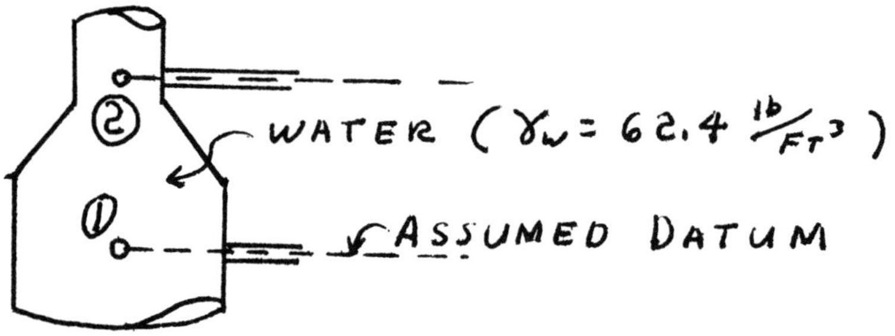

WRITE BERNOULLI'S EQUATION FROM 1 to 2

$$\frac{P_1}{\gamma_w} + Z_1 + \frac{V_1^2}{2g} = \frac{P_2}{\gamma_w} + Z_2 + \frac{V_2^2}{2g}$$

KNOW $Z_1 = 0$,
$Z_2 = 1.5'$

WRITE CONTINUITY EQUATION FROM 1 to 2

$$Q_1 = Q_2 \qquad A_2 = \pi r^2 = \pi \left(\frac{.5}{2}\right)^2$$

$$V_1 A_1 = V_2 A_2 \qquad A_1 = \pi \left(\frac{1}{2}\right)^2$$

$$V_1 = \frac{V_2 A_2}{A_1} \quad \text{so} \quad V_1 = V_2 \frac{\pi \left(\frac{.5}{2}\right)^2}{\pi \left(\frac{1}{2}\right)^2}$$

Now know $V_1 = V_2(.25)$

WRITE MANOMETER EQUATION FROM 1 to 2

$$P_1 + \frac{5}{12}'\gamma_w - \frac{2}{12}'(10.06)\gamma_w - \left[1.5' + \frac{3'}{12}\right]\gamma_w = P_2$$

$$P_1 - 187.82\ \frac{lb}{Ft^2} = P_2$$

Now know $P_1 - P_2 = 187.82\ \frac{lb}{Ft^2}$

REWRITE BERNOULLI'S EQUATION

$$\frac{P_1 - P_2}{\gamma_w} - Z_2 = \frac{V_2^2 - V_1^2}{2g}$$

$$\left[\frac{P_1 - P_2}{\gamma_w} - Z_2\right] 2g = V_2^2 - (.25V_2)^2 = .9375\,V_2^2$$

$$\left[\frac{187.82}{62.4} - 1.5\right] 64.4 = .9375\,V_2^2 \Rightarrow V_2^2 = 103.72$$

$$V_2 = 10.18\ \frac{FT}{S}$$

Now $Q = Q_2 = V_2 A_2 = 10.18\left(\pi\left(\frac{.5}{2}\right)^2\right)$

$$Q = 2\ \frac{F+^3}{S}$$

MEASUREMENTS OF COMPRESSIBLE FLOW

13-12

Velocity of a supersonic air flow is to be measured by using a Pitot tube. The static pressure upstream of the Pitot tube is 70 kPa. The total pressure indicated by the Pitot tube is 210 kPa. The total temperature is indicated as 400°K. Determine the velocity of the free stream.

****

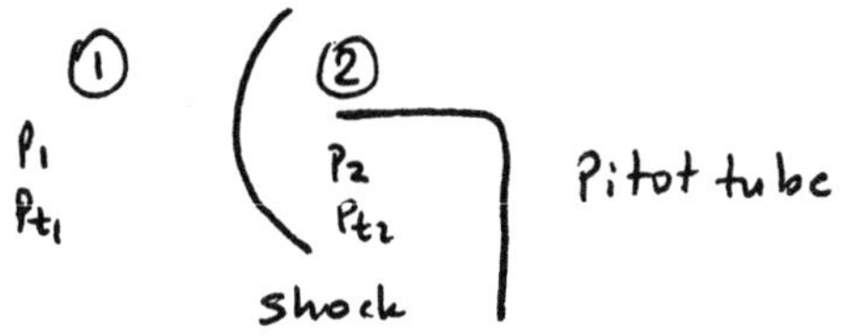

For air, $k = 1.4$ (ratio of specific heats).

Rayleigh's formula :

$$\frac{p_1}{p_{t2}} = \frac{\left\{\frac{2k}{k+1} \cdot M_1^2 - \frac{k-1}{k+1}\right\}^{\frac{1}{k-1}}}{\left(\frac{k+1}{2} \cdot M_1^2\right)^{\frac{k}{k-1}}}$$

after substituting the given values, we get

$$\frac{70}{210} = \frac{\left\{\frac{2.8}{2.4} M_1^2 - \frac{0.4}{2.4}\right\}^{2.5}}{\left(\frac{2.4}{2} M_1^2\right)^{3.5}}$$

This equation can be solved in several ways. Using an iterative type solution procedure, we can get $M_1 \cong 2.3298$.

Static temperature before the shock wave :

$$\frac{T_{t1}}{T_1} = \left(1 + \frac{k-1}{2} M_1^2\right) \rightarrow T_1 = \frac{T_{t1}}{(1 + 0.2\, M_1^2)} = \frac{400}{(1 + 0.2 \times 2.3298^2)}$$

$$T_1 = 191.79 \text{ °K}$$

Speed of sound in front of the shock wave :

$$c = \sqrt{k.R.T} = \sqrt{1.4 \times 287 \times 191.79} = 277.6 \text{ m/s}$$

$\therefore$ free stream velocity $M_1 = \frac{V_1}{c_1} \rightarrow V_1 = M_1 . c_1 = 2.3298 \times 277.6$

$$= 646.75 \text{ m/s}.$$

14

TURBOMACHINERY

PUMPS

14-1

A compressor is used to increase the pressure of oxygen (gas) from 20°C and 80 kPa to 200 kPa. Flow rate of the oxgen gas is 3 kg/s. If the efficiency of this compressor is 0.70, determine the actual power required by the compressor.

****

Assuming that the process in the compressor is adiabatic, energy eqn. gives (for unit mass) :

$$W_c = h_2 - h_1 = C_p (T_2 - T_1) \qquad \text{(oxygen is assumed as ideal gas)}$$

$$\therefore \quad W_c = C_p T_1 \left(\frac{T_2}{T_1} - 1 \right)$$

For ideal compressor, compression process is isentropic and we can write :

$$\frac{T_2}{T_1} = \left(\frac{P_2}{P_1}\right)^{\frac{k-1}{k}}$$

$$\therefore \quad W_c = C_p T_1 \left(\left(\frac{P_2}{P_1}\right)^{\frac{k-1}{k}} - 1 \right) \quad , \quad \text{for an ideal gas } C_p = \frac{kR}{k-1}$$

$$\therefore \quad W_c = \frac{kRT_1}{k-1} \left(\left(\frac{P_2}{P_1}\right)^{\frac{k-1}{k}} - 1 \right)$$

Total ideal work:

$$\dot{W}_c = \dot{m}\, w_c = \dot{m}\, \frac{k R T_1}{k-1}\left(\left(\frac{P_2}{P_1}\right)^{\frac{k-1}{k}} - 1\right)$$

$$= 3 \times \frac{1.4 \times 8314.3 \times 293}{(1.4-1) \times 32.0}\left(\left(\frac{200}{80}\right)^{\frac{0.4}{1.4}} - 1\right)$$

$$= 239\,802.6\ \frac{\text{lbf-ft}}{\text{s}} = \frac{239802.6}{550} = 436\ \text{hp}$$

Actual work required by the compressor:

$$(\dot{W}_c)_{ACT.} = \frac{\dot{W}_c}{\eta_c} = \frac{436}{0.7} = 622.8\ \text{hp}$$

SPECIFIC SPEED

14-2

Determine the specific speed of a pump which would deliver 1,600 gpm against a head of 150 ft while operating at a speed of 1,800 rpm.

****

Discharge of pump in gallons per minute = 1,600 gpm

Head = 150 feet, Speed in revolutions per minute = 1,800 rpm

$$\text{Specific Speed} = N_s = \frac{n_r\,(gpm)^{1/2}}{H^{3/4}}$$

$$\text{or, } N_s = \frac{1800 \times \sqrt{1600}}{(150)^{3/4}} \simeq 1680$$

SUCTION DESIGN

14-3

For the suction system of the pump in the Figure. Calculate the NPSH available for a flow of Q=0.4 ft^3/s of warm water (s.w.=62.0lb/ft^3). The athmospheric pressure at the site is 14 psia. Use f=0.027 and vapor water pressure = 1.0ft.

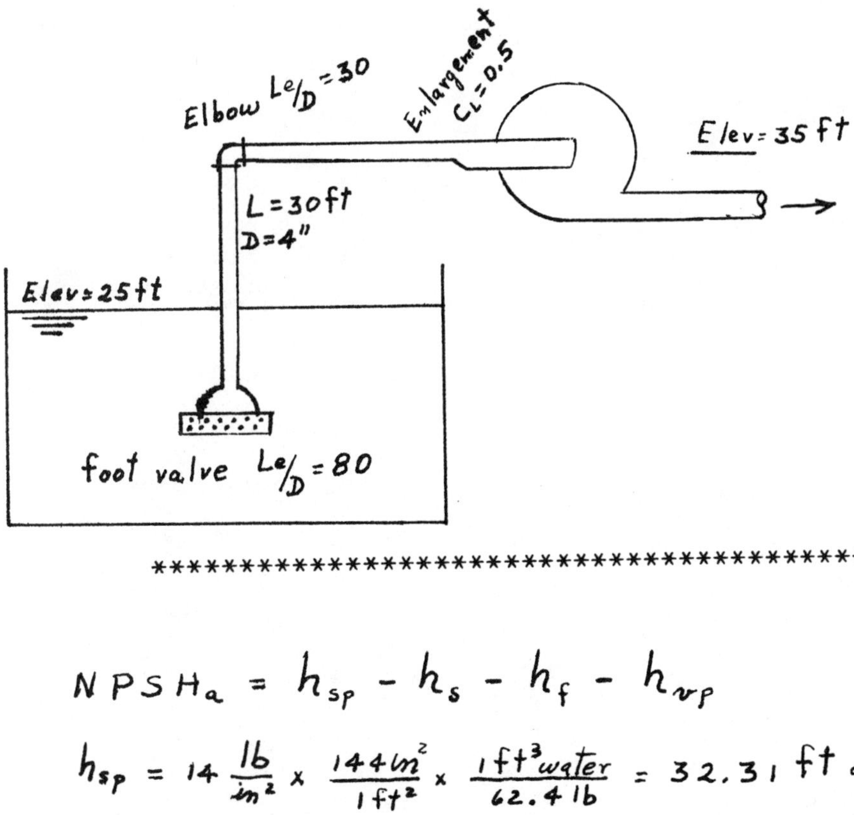

****

$$NPSH_a = h_{sp} - h_s - h_f - h_{vp}$$

$$h_{sp} = 14 \frac{lb}{in^2} \times \frac{144 in^2}{1 ft^2} \times \frac{1 ft^3 water}{62.4 lb} = 32.31 \text{ ft of water}$$

$$h_s = 35 ft - 25 ft = 10 ft$$

$$h_f = h_{pipe} + h_{valve} + h_{elbow} + h_{enlarg.}$$

$$v = \frac{Q}{A} = \frac{0.4 \, ft^3/s}{\pi (1/3 ft)^2/4} = 4.58 \, ft/s$$

$$h_f = f\frac{L}{D}\frac{v^2}{2g} + f\frac{L_e}{D}\frac{v^2}{2g} + f\frac{L_e}{D}\frac{v^2}{2g} + C_L\frac{v^2}{2g}$$

$$= \frac{v^2}{2g}\left[f\left(\frac{L}{D} + \frac{L_e}{D} + \frac{L_e}{D}\right) + C_L\right]$$

$$= \frac{(4.58)^2}{2 \times 32.2}\left[0.027\left(\frac{30}{1/3} + 80 + 30\right) + 0.5\right] = 1.92\ \text{ft}$$

$$h_{vp} = 1.0\ \text{ft}$$

$$NPSH_a = 32.31\ \text{ft} - 10\ \text{ft} - 4.58\ \text{ft} - 1.92\ \text{ft} - 1.0\ \text{ft}$$

$$NPSH_a = 14.81\ \text{ft}$$

15 OPEN CHANNEL FLOW

ENERGY RELATIONS

15-1

Five hundred cubic feet per second flows in a rectangular channel with a width of 12 feet. Using n = 0.015, plot the specific energy curve for depth between 0 and 12 feet. Calculate the critical depth, the minimum specific energy, and the critical slope. With a slope of 0.0002, is the flow sub-critical or super-critical?

****

$$q = Q/b = \frac{500 \text{ cfs}}{12 \text{ ft}} = 41.67 \text{ cfs/ft.}$$

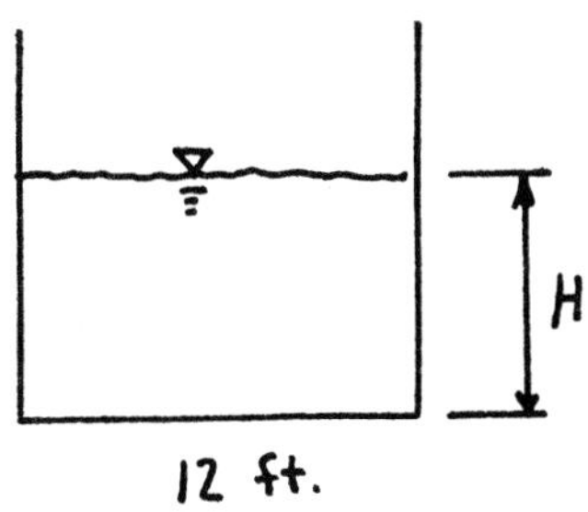

SPECIFIC ENERGY – $E = H + \frac{1}{2g}\left(\frac{q}{H}\right)^2$

H	E	H	E	H	E
1	28.0	5	6.08	9	9.33
2	8.74	6	6.75	10	10.27
3	6.00	7	7.55	11	11.22
4	5.69	8	8.42	12	12.19

CRITICAL SLOPE – S_c

$$S_c = \frac{g m^2}{u^2 y_c^{1/3}}$$

MIN. SPECIFIC ENERGY – E = 5.667

CRITICAL DEPTH = $\frac{2}{3} E_{min}$ = 3.78

$$S_c = \frac{(32.2)(0.015)^2}{(1.49)^2 (3.78)^{1/3}}$$

S_c = 0.0021

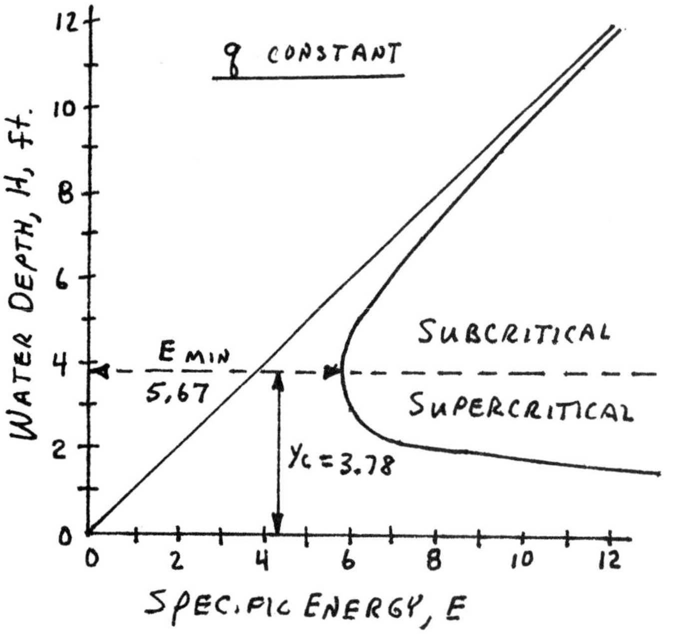

A SLOPE OF

0.0002 < 0.0021

THEREFORE IT IS SUBCRITICAL OR MILD SLOPE.

15-2

Water is flowing in a rectangular channel at a depth of 10 ft and a flow rate of 800 cfs. For the given amount of specific energy in the channel determine the critical depth. The base width of the channel is 20 ft.

****

Discharge per unit width $= q = \frac{Q}{b} = \frac{800}{20} = 40$ cfs/ft

Specific Energy: $E = y + \frac{q^2}{2gy^2} = 10 + \frac{40 \times 40}{2 \times 32.174 \times 100}$

or, $E = 10.2$ ft

Critical depth $= y_c = \frac{2}{3} E = \frac{2}{3}(10.2) = 6.8$ ft

CHANNEL DESIGN

15-3

A rectangular troweled concrete channel (n = 0.013) has a slope of 1 foot in 500 feet and carries water at a rate of 480 cfs. Determine the dimensions of the channel for best hydraulic conditions.

****

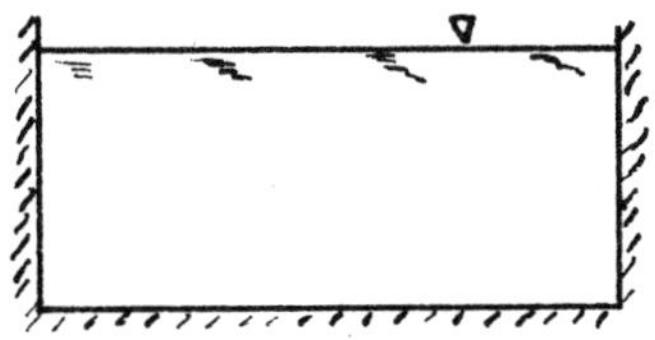

$$R_H = \frac{\text{Area}}{\text{perimeter}} = \frac{bh}{b+2h}$$

for best hydraulic cond: $R_H = \frac{h}{2}$

$$\therefore \frac{h}{2} = \frac{bh}{b+2h} \Rightarrow b = 2h$$

$$Q = \frac{u}{n} A R_H^{2/3} S_o^{1/2}$$

$$480 = \frac{1.49}{.013}(bh)\left(\frac{h}{2}\right)^{2/3}\left(\frac{1}{500}\right)^{1/2} = \frac{1.49}{.013}(2h \times h)\left(\frac{h}{2}\right)^{2/3}\left(\frac{1}{500}\right)^{1/2}$$

$$(h)^{8/3} = \frac{480(.013)(2)^{2/3}}{1.49(2)\left(\frac{1}{500}\right)^{1/2}} = 74.3$$

$$h = (74.3)^{3/8} \Rightarrow \underline{\underline{h = 5.03 \text{ ft}}}$$

$$b = 2h \Rightarrow \underline{\underline{b = 10.06 \text{ ft}}}$$

15-4

Given the rectangular channel in the Fig. is made out of wood boards (n=0.012) at a slope of 0.1%. What is the depth of the water for a flow of Q= 0.8 cfs (Trial and error)

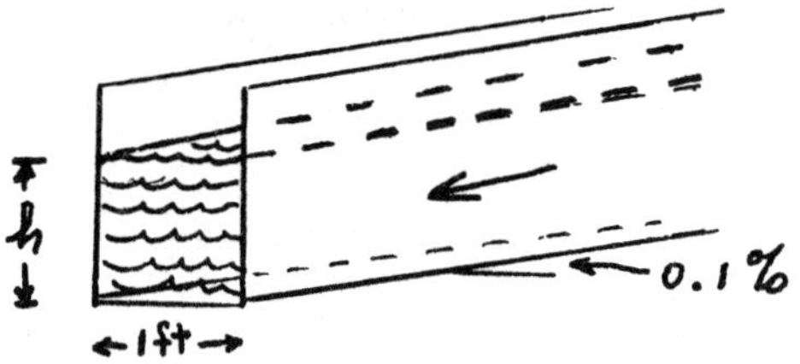

****

$$Q = \frac{1.49}{n} A R^{2/3} S^{1/2}$$

$$A = h * 1ft \qquad R = \frac{A}{P} = \frac{h * 1ft}{2h + 1ft}$$

$$Q = \frac{1.49}{0.012} h \left(\frac{h}{1+2h}\right)^{2/3} (0.001)^{1/2}$$

$$Q = 3.9265\, h \left(\frac{h}{1+2h}\right)^{2/3}$$

By trial and error in my calculator (a programmable one):

h	1ft	0.5	0.6	0.54	0.51
Q	1.887cfs	0.779	0.990	0.862	0.7999

Elevation in the channel: 0.51 ft

15-5

Water is flowing in a rectangular channel at a velocity of 2 m/s. Design the most efficient cross section for a slope of 0.0001. Also, indicate the discharge in the channel. The value of the Manning roughness coefficient is 0.016.

****

Manning Formula: $V = \frac{1}{n} R^{2/3} S^{1/2}$

$$2 = \frac{1}{0.016} R^{2/3} (0.0001)^{1/2}$$

or, $R^{2/3} = 3.2$, $R = 2.172$ m

Hydraulic radius = $R = \frac{A}{P} = \frac{by}{b+2y}$

For an efficient cross section, width = 2 x depth

or, $R = \frac{by}{b+2y} = \frac{2y^2}{4y} = \frac{y}{2} = 2.172$ m

channel depth, $y = 4.344$ m.

channel width = $b = 2y = 8.688$ m

Discharge = $Q = AV = 4.344 \times 8.688 \times 2$

or, $Q = 75.481 \ m^3/s$

UNIFORM FLOW

15-6

Water is flowing in a triangular channel at a flow depth of 3 ft as shown. For a flow rate of 10 cfs determine the slope of the channel. Take the value of the Chezy coefficient to be 150.

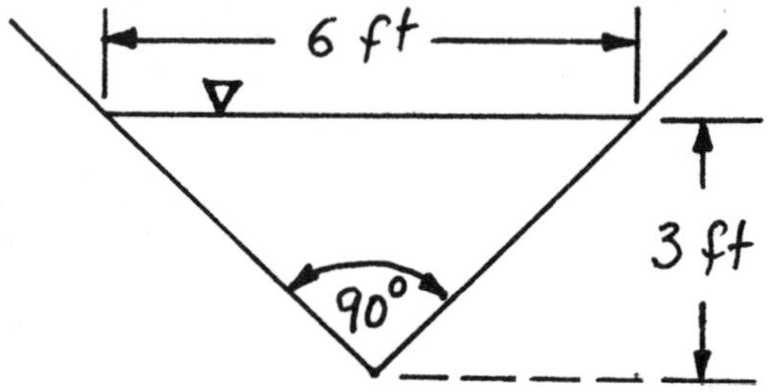

****

Area $= A = y^2 = 9 \text{ ft}^2$, Hydraulic radius $= R$

$$= \frac{A}{P} = \frac{9}{2\sqrt{2}(3)} = 1.061 \text{ ft}$$

velocity $= V = \frac{Q}{A} = \frac{10}{9} = 1.111$ ft/sec

Chezy formula, $V = C\sqrt{RS}$

or, $1.111 = 150\sqrt{1.061}\; S^{1/2}$

$$S^{1/2} = \frac{1.111}{150\sqrt{1.061}}$$

or, Slope $= S = 0.000052$

15-7

A circular culvert under a highway is 5 Ft. in diameter and made of corrugated metal. It drops 2 Ft. over a length of 500 Ft. Calculate the normal discharge when the culvert runs half full.

****

MANNING FORMULA FOR OPEN CHANNEL

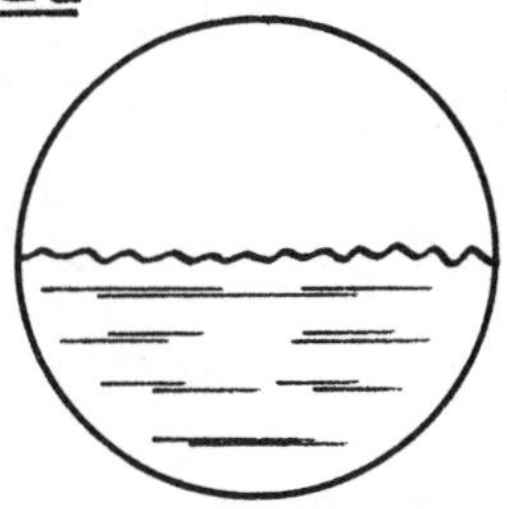

SLOPE $= 2/500$ $\quad n = .026$

$A = \pi(5)^2/4(2)$

$$R_H = A/P = \frac{25\pi/8}{5\pi/2} = 1.25$$

MANNING:

$$Q = \frac{1.49}{.026} A R^{2/3} S^{1/2} = \frac{1.49}{.026}(3.125)(1.25)^{2/3}\left(\frac{1}{250}\right)^{1/2}$$

$Q = 13.14$ CFS

15-8

The open channel shown has a slope of .169% and is made of formed, unfinished concrete. Calculate Q.

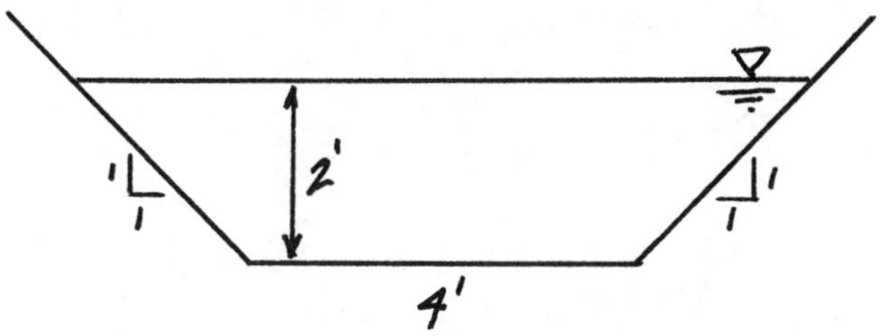

****

Manning's Eqn: $Q = \frac{1.49}{n} A R^{2/3} S^{1/2}$

$A = 4 \times 2 + 2\left(\frac{1}{2}\right)(2)(2) = 12 \text{ FT}^2$

$WP = \frac{2}{.707} \times 2 + 4 = 9.66'$

$R = \frac{A}{WP} = \frac{12}{9.66} = 1.24$

$n = .017$ (table)

$Q = \frac{1.49}{.017} (12)(1.24)^{2/3}(.00169)^{1/2} = \boxed{49.9 \text{ Ft}^3/\text{s}}$

15-9

The stream cross section shown below is constructed of smooth asphalt with a bottom slope of 0.001. Determine the uniform rate of flow in cfs.

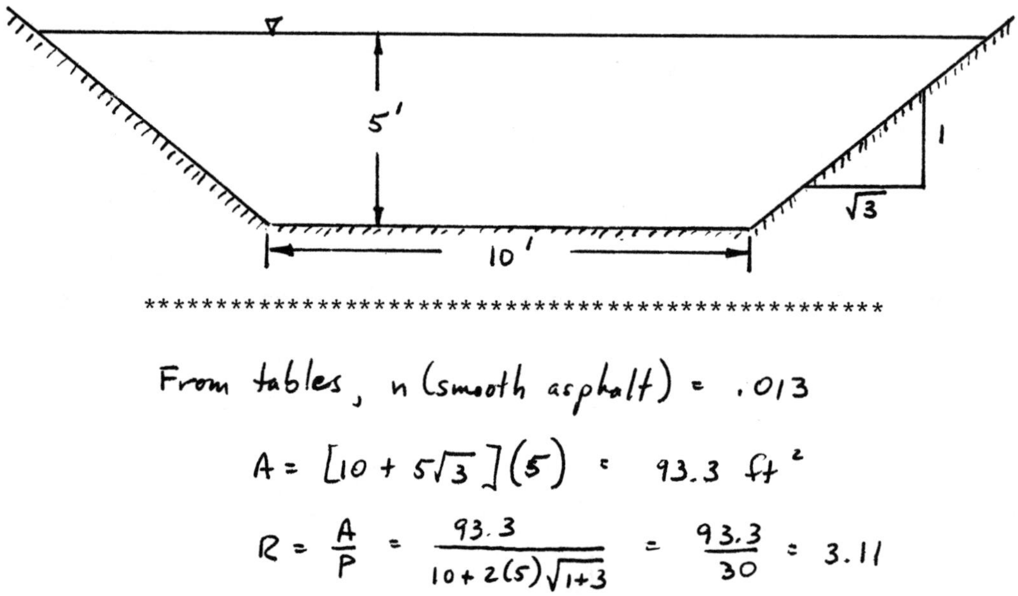

****

From tables, n (smooth asphalt) = .013

$A = [10 + 5\sqrt{3}](5) = 93.3 \text{ ft}^2$

$R = \frac{A}{P} = \frac{93.3}{10 + 2(5)\sqrt{1+3}} = \frac{93.3}{30} = 3.11$

Manning Equation: $Q = \frac{1.49}{n} A R^{2/3} S^{1/2}$

$Q = \frac{1.49}{.013} (93.3)(3.11)^{2/3}(.001)^{1/2} = \underline{\underline{720 \text{ cfs}}}$

15-10

Water is flowing in a 16 inch diameter sewer pipe at a depth of 5 inches. The pipe has a slope of 0.25 inch per foot. Determine the velocity of the flow and flowrate.

****

ASSUME HAZEN-WILLIAM'S COEF. – $C_{HW} = 100$

$$V = 1.318\, C_{HW}\, R_h^{0.63}\, S^{0.54}$$

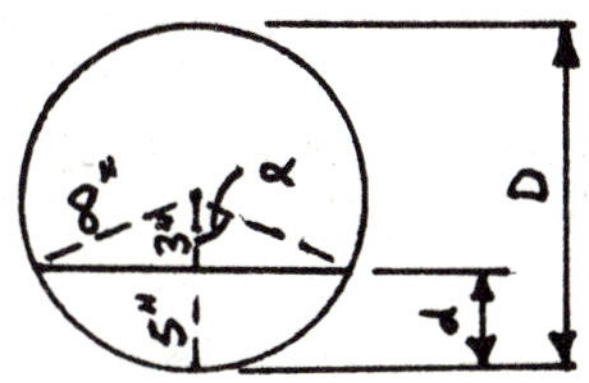

USE PARTIAL FLOW DIAGRAM, IF AVAILABLE, TO DETERMINE R_h OR COMPUTE AS FOLLOWS.

$$\alpha = \cos^{-1}\frac{3}{8} = 67.98°$$

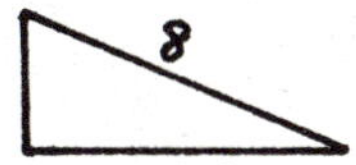

$$A = \frac{(3)(7.416)}{(2)(144)} = 0.0773\ ft^2$$

$$A = \left(\frac{\pi D^2}{4}\right)\frac{2\alpha}{360} - (2)(0.0773) = \frac{(3.14)\left(\frac{16}{12}\right)^2}{4}\,\frac{2(67.98)}{360} - (2)(0.0773)$$

$$A = 0.373\ ft^2$$

$$P = \pi D\,\frac{2\alpha}{360} = (3.14)\left(\frac{16}{12}\right)\frac{(2)(67.98)}{360} = 1.582$$

$$R_h = \frac{A}{P} = \frac{0.373}{1.582} = 0.236 \qquad S = \frac{0.25\ in.}{12\ in.} = 0.0208$$

$$V = 1.318\,(100)\,(0.236)^{0.63}\,(0.0208)^{0.54} = \underline{\underline{6.56\ fps}}$$

$$Q = AV = (0.373\ ft^2)(6.56\ fps) = \underline{\underline{2.45\ cfs}}$$

15-11

A rectangular, concrete lined channel (n=0.015) will have a drop of five feet per mile. It is required that the channel be able to carry 150 cfs of water when flowing full. Find the minimum required channel dimensions if the width is to be twice the depth.

****

MANNINGS EQUATION: $V = \frac{1.49}{n} R^{2/3} S^{1/2}$

So $Q = 150\ ft^3/sec = AV = \frac{1.49}{n} A R^{2/3} S^{1/2}$

$S = \text{slope} = \frac{5'}{5280'} = 0.000947$

For

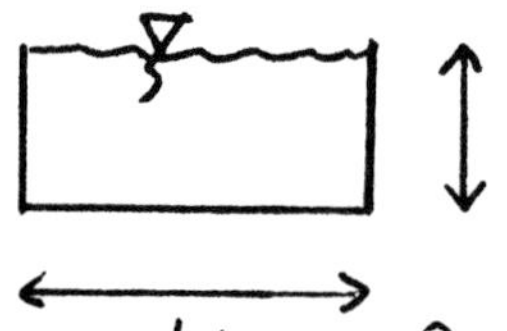

depth = d

width = 2d

AREA = $d \cdot 2d = 2d^2$

WETTED PERIMETER = $d + d + 2d = 4d$

Hydraulic Radius $= \frac{A}{P} = \frac{2d^2}{4d} = \frac{d}{2}$

So $150 = \frac{1.49}{0.015} \left(\frac{d}{2}\right)^{2/3} (.000947)^{1/2} (2d^2)$

$150 = 3.85\, d^{8/3} \Rightarrow d = \left(\frac{150}{3.85}\right)^{3/8}$

$d = 3.95'$

$b = 2d = 7.90'$

15-12

Uniform flow occurs in a concrete (gunite) channel with a base of 10 feet and side wall slopes of 2 horizontal to 1 vertical. The channel is sloped one foot in a 1000 feet. With a water depth of 5 feet determine the velocity and flowrate of the water.

****

USE CHEZY - MANNING EQUATION – $Q = \frac{1.49}{m} A R_h^{2/3} S_o^{1/2}$

$m = 0.019$

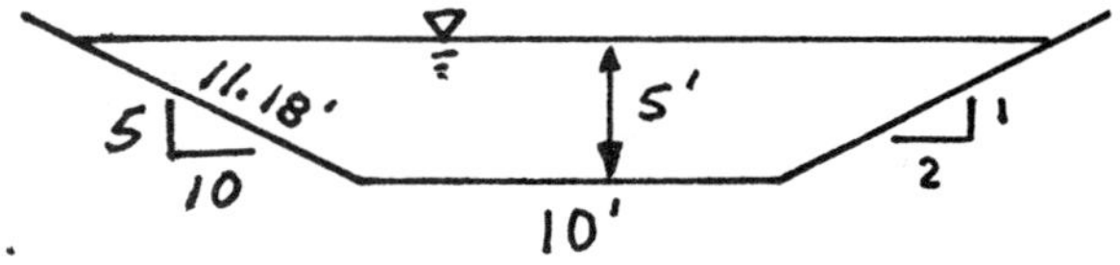

$A = 20 \times 5 = 100\ ft^2$

$P = 11.18 + 10 + 11.18 = 32.36\ ft.$

$R_h = \frac{A}{P} = \frac{100}{32.36} = 3.09\ ft.$

$S_o = \frac{1}{1000} = 0.001$

$$Q = \frac{1.49}{0.019}(100)(3.09)^{2/3}(.001)^{1/2} = \underline{\underline{526\ cfs}}$$

$$V = \frac{Q}{A} = \frac{526\ cfs}{100\ ft^2} = \underline{\underline{5.26\ fps}}$$

15-13

Water is flowing in a wide rectangular channel at a depth of 2 m. Determine the velocity in the channel if the slope is 0.000 225. Consider the Manning roughness coefficient to be 0.035.

****

Hydraulic radius $= R = y = 2\ m$

Manning Formula, $V = \frac{1}{n} R^{2/3} S^{1/2}$

or, $V = \frac{1}{0.035}(2)^{2/3}(0.000225)^{1/2} = 0.680\ m/s$

15-14

An open channel of 0.2% slope and finished in concrete (n=0.015) is running at a steady flow 1.5 ft deep as indicated in the Figure. What is the flow of water in this channel? Use the Manning formula.

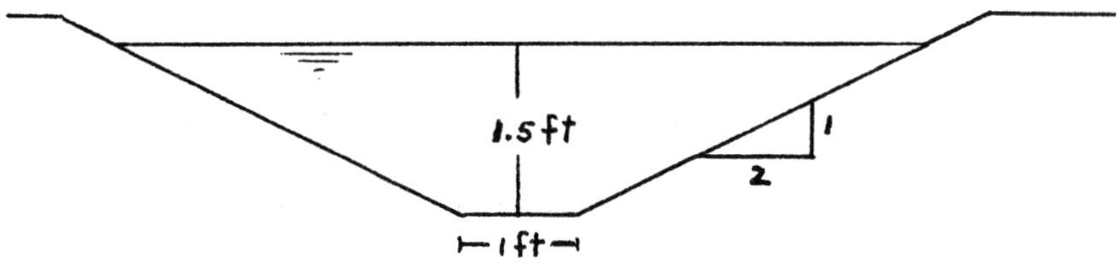

****

$$Q = \frac{1.49}{n} A R^{2/3} S^{1/2}$$

$$A = (1\,ft \times 1.5\,ft) + (1.5\,ft \times 3\,ft) = 5\,ft^2$$

$$S = 0.002\ ft/ft$$

$$R = \frac{A}{\text{Wet Perim.}} \qquad \text{Wet Perim} = 1\,ft + 2\sqrt{(1.5\,ft)^2 + (3\,ft)^2}$$

$$= 7.71\,ft$$

$$= \frac{5\,ft^2}{7.71\,ft} = 0.65\,ft$$

$$Q = \frac{1.49}{0.015}\ 5 \times (0.65)^{2/3} \times (0.002)^{1/2}$$

$$= 16.64\ ft^3/sec$$

HYDRAULIC JUMP

15-15

A hydraulic jump occurs in a rectangular channel. The water depth upstream from the jump is 3 feet and water depth downstream from the jump is 8 feet. Calculate the critical depth.

****

$$\frac{y_2}{y_1} = \frac{1}{2}\left[-1 + \sqrt{1 + \frac{8q^2}{g y_1^3}}\right]$$

$$\frac{8}{3} = \frac{1}{2}\left[-1 + \sqrt{1 + \frac{8q^2}{(32.2)(3)^3}}\right] \qquad q = 65.2 \text{ cfs/ft}$$

$$y_c = \sqrt[3]{q^2/g} = \sqrt[3]{\frac{(65.2)^2}{32.2}} = \underline{\underline{5.08 \text{ ft}}}$$

15-16

A hydraulic jump is observed to occur in a channel with a mild slope at a flow rate of 15 cfs/ft. If the depth before the jump is 1 ft, determine the depth after the jump.

****

The hydraulic jump formula is:

$$y_2 = \frac{y_1}{2}\left(-1 + \sqrt{1 + \frac{8q^2}{g y_1^3}}\right)$$

or, $$y_2 = \frac{1}{2}\left(-1 + \sqrt{1 + \frac{8 \times 15 \times 15}{32.174 \times (1)^3}}\right)$$

or, $y_2 = 3.27$ ft

15-17

Water flows from beneath a sluice gate into a horizontal channel and forms an hydraulic jump as shown below. The losses through the gate are 5 m. Assume the reservoir velocity head negligible. Using the specific energy diagram provided, determine the depth of flow downstream of the jump and the energy loss through the jump.

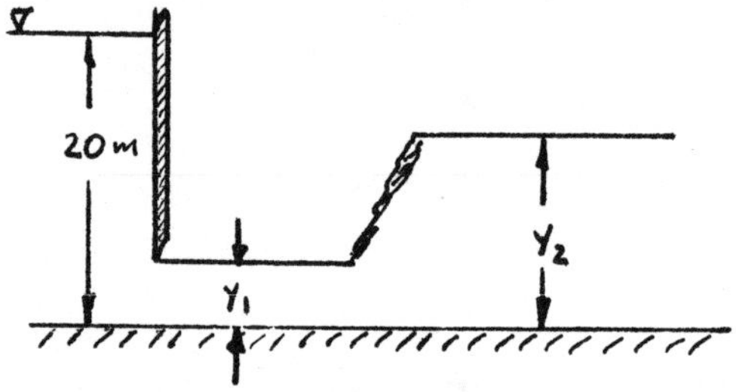

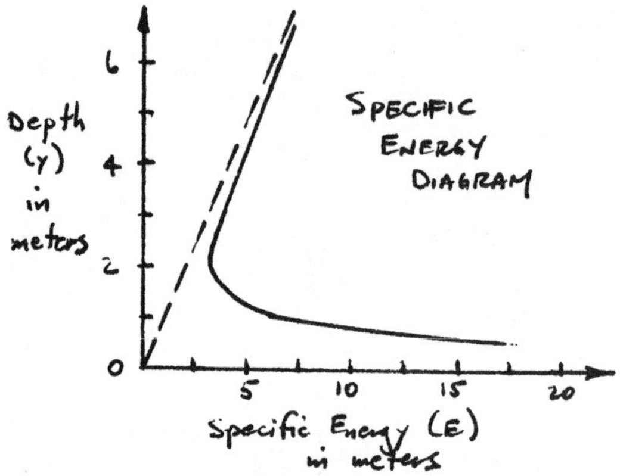

****

$$E_0 = y + \frac{V^2}{2g}^{\,0} = 20 \text{ m}$$

$$E_1 = 20 - 5 = 15 \text{ m} = Y_1 + \frac{V_1^2}{2g}$$

From Specific Energy diagram : $Y_1 = 0.6$ m @ $E_1 = 15$ m

$$\therefore \; V_1 = \sqrt{(15 - .6) * 19.6} = 16.8 \text{ mps}$$

Hydraulic jump equation :

$$Y_2 = \frac{Y_1}{2}\left[-1 + \sqrt{1 + 8\mathbb{F}_{r_1}^2}\right]$$

$$\mathbb{F}_{r_1}^2 = \frac{V_1^2}{gy_1} = \frac{(16.8)^2}{9.8(0.6)} = 48$$

$$Y_2 = \frac{0.6}{2}\left[-1 + \sqrt{1 + 8(48)}\right] = \underline{\underline{5.6 \text{ m}}}$$

From S.E. diagram, $E_2 = 6$ m @ $y_2 = 5.6$ m

$$h_j = E_1 - E_2 = 15 - 6 = \underline{9 \text{ m}}$$